PADDLING THE PASCAGOULA

PADDLING THE

PASCAGOULA

ERNEST HERNDON AND SCOTT B. WILLIAMS

UNIVERSITY PRESS OF MISSISSIPPI / JACKSON

Publication of this book was made possible in part by Cathy Shropshire.

www.upress.state.ms.us
www.scottbwilliams.com
www.ernestherndon.com

The University Press of Mississippi is a member of the Association of American University Presses.

Photographs by Ernest Herndon and Scott B. Williams unless otherwise noted

Manufactured in the United States of America

First edition 2005
∞

Library of Congress Cataloging-in-Publication Data
Herndon, Ernest.
Paddling the Pascagoula / Ernest Herndon and Scott B. Williams.— 1st ed.
p. cm.
Includes bibliographical references.
ISBN 1-57806-714-6 (pbk. : alk. paper)
1. Herndon, Ernest—Travel—Mississippi—Pascagoula River. 2. Williams, Scott B.—Travel—Mississippi—Pascagoula River. 3. Canoes and canoeing—Mississippi—Pascagoula River. 4. Sea kayaking—Mississippi—Pascagoula River. 5. Natural history—Mississippi—Pascagoula River. 6. Pascagoula River Region (Miss.)—Description and travel. 7. Pascagoula River (Miss.)—Description and travel. 8. Pascagoula River (Miss.)—Environmental conditions. I. Williams, Scott B. II. Title.
F347.P26H47 2005
917.62′12—dc22 2005006830

British Library Cataloging-in-Publication Data available

This book is dedicated to conservationist Cathy Shropshire, who got us launched, and editor Craig Gill, who saw us through.

CONTENTS

IX FOREWORD

3 PART ONE
LEAF RIVER
Ernest Herndon

79 PART TWO
CHICKASAWHAY RIVER
Scott B. Williams

153 PART THREE
UPPER PASCAGOULA RIVER
Ernest Herndon

175 PART FOUR
LOWER PASCAGOULA RIVER
Scott B. Williams

194 AFTERWORD
Ernest Herndon

196 BIBLIOGRAPHY

FOREWORD

For far too long the Pascagoula River has lain in obscurity, with only a few intrepid souls experiencing the riches that the river and the surrounding lands can offer to both body and mind. Recently the spotlight has been turned on this Mississippi treasure, and Ernest Herndon and Scott B. Williams add to the illumination with this book's description of their journey.

My own experiences along the Pascagoula have left me with an appreciation of the uniqueness of the region and with pride that Mississippi can lay claim to this extraordinary land. I count myself among the privileged who have held a gulf sturgeon—a prehistoric creature that has continued to find sanctuary in the Pascagoula River. I have searched for black bears in the Leaf River Refuge and watched swallow-tailed kites soar overhead. I have glimpsed at least a small portion of the marvels of this countryside, which remains remarkably intact.

We owe much to the Mississippi Game and Fish Commission, especially Bill Quisenberry, as well as to Graham Wisner and to the Nature Conservancy for their vision and untiring efforts to protect a part of the Pascagoula River in the 1970s, and to Herman Murrah, who first shone the light on this hidden treasure. Their tireless efforts assured that we today could experience the Pascagoula in all its wonder. But continuing vigilance is needed to guarantee that the Pascagoula River remains the gem that it is—that sturgeons continue to return to its waters to spawn, that kites nest in the trees overhead, and that canoes and kayaks can travel the river unimpeded from its headwaters to the Gulf Coast.

This book is another in what I hope will become a wealth of writings about Mississippi's natural resources and how her people interact with and are shaped by them—lessons learned, bonds made, and feelings of optimism that future generations will explore and protect those resources. The Pascagoula is just one of Mississippi's treasures. Others abound, from the Mississippi River to the creek behind the house, from the Delta National Forest to the woodlot where you watch for the first spring trillium. I urge you to plan your own adventure, whether it be a trip on the Pascagoula River or a walk along a quiet trail. Experience the richness that is here. But for now, settle back and share the view from the bow of Ernest's and Scott's vessels as they paddle the Pascagoula.

—DR. CATHY SHROPSHIRE

PADDLING THE PASCAGOULA

Cypress swamps like this are part of the appeal of the Pascagoula basin.

PART ONE

LEAF RIVER

ERNEST HERNDON

Getting ready to go are, from left, Scott Williams, Travis Easley, Eddie McCalip, Ernest Herndon. (photo by Michelle Calvert)

1

I hatched the idea of paddling the entire Pascagoula River, from headwaters to coast, back in the 1990s. (I want to establish that up front in case my coauthor, Scott Williams, tries to claim credit.) During many years of canoeing Mississippi streams, it became apparent to me that the Pascagoula basin was supreme. The 81-mile-long river and its 750 miles of 11 canoeable tributaries offer not only the best paddling in the state but some of the best anywhere, in my opinion. There's no big whitewater, but there are lots of riffles, small rapids, and even some waterfalls. As I marveled at these streams with their abundance of woods and wildlife, I thought: wouldn't it be great to canoe the Pascagoula from head to mouth, from hills to coast, from spring branch to saltwater?

The Pascagoula (Pas-ca-GOO-la) has two major tributaries, the 180-mile Leaf and the 159-mile Chickasawhay (Chick-a-sah-HAY). They run together to form the main Pascagoula River, which empties into the Gulf of Mexico between the towns of Pascagoula and Gautier (GO-shay). My idea was to choose one of these tributaries, start at the highest navigable point, paddle to the sea, and then write a book about it. I contacted University Press of Mississippi, pointing out the fact that the Pascagoula is the last unaltered major river system in the lower 48 states, and the editors expressed interest.

Then I started to do the math. Such a trip would take roughly two weeks and range from 210 miles if I took the Leaf (not all of which is navigable) to 240 via the Chickasawhay. A typical nonfiction book contains some 60,000 words, which means I would have to generate more than 4,000 words per day, or 250 words per mile. As a newspaper outdoors columnist who typically produces, say, a 1,000-word article from an entire weekend float trip, I began to get nervous. I woke up at night going over the calculations. I broke out in a sweat.

The fact that other river books usually detail far longer trips didn't help matters. Kim Trevathan's *Paddling the Tennessee River* was based on 652 miles and five weeks. For his book *Mississippi Solo*, Eddy L. Harris canoed the entire Mississippi River. Johnny Molloy took two months to float the Suwannee River and Florida's west coast in *From the Swamp to the Keys.* W. Hodding Carter went up the Missouri River in a motorized dinghy for *Westward Whoa!*, then traveled from Greenland to Canada in an open boat for *A Viking Voyage*. In *River Horse*, William Least Heat-Moon spent five months journeying 5,222 miles in a yacht. The only book I know of based on a short float was Franklin Burroughs's *The River Home: A Return to the Carolina Low Country*, an account of a six-day sojourn on the Waccamaw River. Burroughs used the experience as a springboard to explore in-depth history such as his ancestors' Civil War journals, devoting the entire first chapter to the origins of Horry County, South Carolina. Moreover, as an English professor, Burroughs writes with microscopic detail and many fine metaphors. I appreciate a well-turned phrase, but I'm a journalist who churns out newspaper stories to a daily deadline and generally lives by Mark Twain's adage, "If you see an adjective, kill it!"

Panic set in. Then I had an inspiration. Since the Pascagoula has two major tributaries of more or less equal length, what if I got my buddy Scott Williams to coauthor the book with me? He could do one tributary and I the other. Scott and I have been paddling partners for years, and I also knew him to be an excellent writer, with credits in magazines like *Sea Kayaker* and *Canoe & Kayak*, a guidebook to the Mississippi Gulf Coast, and a nonfiction book on his solo sea kayak expedition across the Caribbean. A coauthorship would mean just 30,000 words each. Plus we'd be covering the Pascagoula basin more thoroughly. Now that sounded doable!

Scott jumped at the idea. He would sea kayak one tributary while I canoed the other. We would meet at the juncture and finish the trip together, side by side in our respective vessels. Sounded almost heroic, a scene requiring the swelling notes of an orchestra, John Philip Sousa perhaps.

Then drawbacks began to occur to me. Suppose—just suppose—the book climbed its way up the *New York Times* best seller list? Got nominated for a National Book Award, the Pulitzer, the Nobel, Oprah Winfrey's Book Club? Made six- or seven-figure profits? Granted, all that seems unlikely for a work on an obscure river in an overlooked state, but look at what James Joyce did with *Finnegans Wake*, a book about the rivers of Ireland. Or Joseph Conrad with *Heart of Darkness*, about a float trip in Africa.

Or Herman Melville and that whale story. In the world of publishing, anything is possible. If the Pascagoula book hit the big time, I would have to share the limelight with Scott—and it was my idea! I began to feel like one of the characters in B. Traven's *Treasure of the Sierra Madre*, pals who strike gold and turn into instant enemies with guns drawn. But we hadn't struck gold yet, so laziness trumped ambition as I decided less work was preferable to more glory.

Writing aside, I was excited about the trip. If there is one river in the United States worth seeing from top to bottom, it's the Pascagoula. This southeast Mississippi waterway is the last major river system in the continental United States whose natural flow is unaltered by dams, channelization, levees, or other human impact. This was documented in the November 4, 1994, issue of *Science* magazine. Authors Mats Dynesius and Christer Nilsson surveyed all the large river systems in the northern part of the world. They found that only 39 percent of those in the Northern Hemisphere are not affected by dams and channelization, and nearly all of those are in the extreme north.

In North America north of Mexico, only 35 such natural river systems remain, all of them in Alaska and northern Canada except one. That one is the Pascagoula. Other unaffected rivers in North America include such far-north streams as the Stikine, Yukon, Noatak, and Back. Moderately affected rivers include the Skagit, Fraser, Mackenzie, Hudson, Delaware, Susquehanna, Potomac, and Pee Dee. Numbered among the strongly affected river systems are the Colorado, Sacramento, Klamath, Columbia, Churchill, St. Lawrence, St. John, Penobscot, Kennebec, Santee, Apalachicola, and Mississippi. That makes the Pascagoula special indeed.

"Large areas in the northern third of the world completely lack unregulated LRSs [large river systems]," says the article, entitled "Fragmentation and Flow Regulation of River Systems in the Northern Third of the World." "Although river exploitation may have different effects in different rivers, some inevitable consequences stand out. For example, several types of important habitats, such as waterfalls, rapids, and floodplain wetlands, may disappear from entire regions. The loss of waterfalls and rapids indicates the loss of numerous species of plants and animals specific to running waters. . . . As a result of habitat destruction and obstruction to organism dispersal, many riverine species may have become extinct over vast areas, whereas populations of others have become fragmented and run the risk of future extinctions."

A color map which accompanies the article illustrates the situation vividly. The color red represents river systems whose flow has been strongly affected. Green covers the unaffected river systems. The lower 48 states are almost completely blood-red. But there's a plume of green in the Deep South—where the Pascagoula flows. This ecological holdout is partly because Mississippi is one of the poorest and least developed states. But it's also because of some astute and far-sighted political maneuvering in the 1970s which resulted in an environmental coup: the setting aside of much of the land bordering the Pascagoula in state wildlife management areas. Donald Schueler's 1980 book *Preserving the Pascagoula* tells the complex story behind the groundbreaking accomplishment involving the Nature Conservancy. As a result of this achievement, the Pascagoula is now bordered by the 37,124-acre Pascagoula River Wildlife Management Area and the 9,494-acre Ward Bayou WMA. It's also flanked by other impressive swaths of public land: the 501,000-acre De Soto National Forest, 940-acre Leaf Wilderness Area, 5,000-acre Charles M. Deaton and Herman Murrah Nature Conservancy Preserves, 19,273-acre Mississippi Sandhill Crane National Wildlife Refuge, 18,500 Grand Bay Reserve, and 7,000-acre Grand Bay NWR—roughly 600,000 acres of wildlands.

As for canoeing, there are opportunities aplenty. The Chickasawhay absorbs Chunky River and Buckatunna (Buck-a-TUN-na) Creek. Into the Leaf run the Okatoma (O-ka-TO-ma), Bowie (also spelled Bouie), Tallahala (Tal-a-HAY-la) and Bogue Homa (Bo-ga HO-ma). The Pascagoula receives Black Creek, Red Creek, and Escatawpa (Es-ca-TAW-pa) River. All these provide excellent, slow-water, deep-woods paddling, with occasional rapids to liven things up. Is it any wonder Scott and I were eager to go?

When it came time to decide who would float which tributary, I magnanimously told Scott it made no difference to me whatsoever. However, given a choice, I noted that I'd prefer the Leaf for the simple reason that I had canoed less of it than I had the Chickasawhay.

Naturally, therefore, Scott wanted the Leaf. That way, he said, he'd be able to discourse at length about his lifelong connection to that side of the

Pascagoula basin. His grandmother's farm was located on Bowie Creek, and Scott grew up playing in the stream. He's also floated it and other Leaf tributaries (of course, so have I). Thus, for him the pressure of delivering 30,000 words from a two-week trip would be softened by his ability at any time to launch into a flashback. He'd probably be able to work his pioneer ancestors in somehow.

Readers were spared the prospect of such endless anecdotes when Scott got a chance to make extra money for floating the Chickasawhay. (So much for literature being more sacred than lucre.) Chris Bowen, head of the Pat Harrison Waterway District, heard about our trip and asked me if we'd be willing to survey log obstructions on the Chickasawhay. The job would involve extensive use of a Global Positioning System and digital camera to check out well over 100 snags identified in a previous survey. As a full-time newspaper reporter and part-time freelance writer, I had too much going on to undertake such a task. But I mentioned the offer to Scott who, being self-employed and experienced in survey work, liked the idea. So now that more money entered the picture, he wanted to float the Chickasawhay; to heck with reminiscing about old times on the Leaf.

Being the good friend and advisor that I am, I pointed out to Scott that it would be grossly unethical for him to receive two sources of income for the same trip. To prevent such conflicts I suggested he give me his part of the book advance. I also noted that I had already done considerable research on the Chickasawhay and expected to be remunerated accordingly. Coincidentally, the value of my efforts came to almost the exact amount the Pat Harrison Waterway District was going to pay him.

Predictably, Scott was unimpressed with either my ethical concerns or the hard work I'd done in good faith. "What does writing have to do with ethics?" he retorted in one of the numerous e-mails we sent back and forth during this planning period. Nor did he care about my research, because the Pat Harrison folks were going to give him tons of data on snags, from which he could get many thousands of words. (I can't wait to read his section on the Chickasawhay, can you?) So, by sheer good fortune, I would get to float the Leaf after all, just as I had originally hoped.

Now I had to find someone to go with me. Unlike Scott, who has made long trips solo—sea kayaking the Caribbean, then the Mississippi River—I prefer human company in the wilderness. Someone to bounce ideas and observations off of. Someone to discuss profound topics with around the campfire. Someone to coerce into sharing the camp chores.

Other than Scott, I had two main paddling buddies, Eddie McCalip and Steve Cox, both of McComb. Eddie, a cameraman for a Mississippi nature show, was suffering from a shoulder injury and couldn't make the trip. Steve, an architect, typically goes with me on long-weekend floats, and his response to my invitation on this one was, "Two weeks! I can't be gone two weeks!"

Next I turned to John Ruskey, who owns Quapaw Canoe Company in Clarksdale, Mississippi, and leads trips on the Mississippi River. He and I had floated the Yazoo River together one time and I found him to be quite companionable. He was interested in the trip, but he already had clients booked for April, when we planned to go.

My list of candidates was dwindling fast. I decided to change tactics. Forget those middle-aged types, bowed down by business or health concerns. I needed somebody young, bold, vigorous; somebody with vision; somebody not enslaved by the workaday world; somebody who hadn't been with me on a long campout. I remembered Travis Easley, a 29-year-old Amite County environmentalist logger. I'd met this interesting young man at a public meeting on scenic streams and we hit it off, especially since he was an avid paddler. He later proved his mettle on an overnight canoe trip in northern Louisiana when we almost wound up at the bottom of Black Bayou.

On that trip, Travis and I had forced our canoe through a thicket of cypress and willow trees trying to get to upper Black Bayou Lake outside Monroe, Louisiana, in rainy 41-degree weather. We finally reached open water, but far from being a lake, it was just a pool hemmed in by marsh grass. We found a passage into another opening, and another, and thus picked our way through a bewildering conglomeration of ponds and channels until we could go no farther. I stood up in the stern and spied what might be the actual lake past a couple hundred yards of grass. There was something funny about this grass, though. When I shoved my paddle against it, the blade plunged through to deep water below. The grass was a floating mattress. If a person were to step on it, he'd find himself over his head in icy water—and no way to get to solid ground. We decided to retreat to our starting point, but getting to shore wasn't easy. Our route had been so roundabout that there was little sense in retracing it. Instead, we figured we'd muscle our way directly to the bank, over floating grass and through willow thickets. Fortunately, Travis's line of work gives him a strong back and arms. We paddled, poled, pushed, and pulled—did everything except step out onto the grass and invite hypothermia. At last we reached solid ground, relieved to have escaped the clutches of Black Bayou. Travis had remained calm throughout—I hoped that was an indicator of what I could expect from him if we ran into problems on the Pascagoula.

When I invited him, he responded eagerly. It turned out he had been dreaming about a Pascagoula float and had already amassed maps. I had a companion. Now I had to choose a boat out of the several battered specimens I own. The choice wasn't easy.

In 1991 Scott had called me to say he'd seen a couple of canoes at Hudson's Salvage Center in Hattiesburg, Mississippi—quality models but badly damaged. At the time I owned a banged-up secondhand aluminum canoe, so I was in the market for an upgrade. I drove to Hattiesburg to take a look. They were quality all right, a red Old Town Canadienne and a pale-blue Sawyer, both 17 feet long, high-grade fiberglass, very sleek. But they'd apparently fallen off an 18-wheeler. The gunwales were bent, the fiberglass cracked and punctured. The original retail sticker prices, still in place, showed each boat would have sold for around $1,000. Hudson's was asking $150 each. I did a bit of haggling and got the price down to $250 for the pair. Then I dropped the canoes off at Scott's house—he wasn't home—with a note reading, "Fix them up and keep the Sawyer." Before long I had a bright red Old Town—patched, battered, slightly warped, but by far the best canoe I'd ever owned. With its sharp entry lines and long, straight hull, it moved swiftly and tracked well. But I discovered it was slow to turn, a drawback when you have to maneuver in swift water. I realized that, as its name suggested, it was better suited for windswept Canadian lakes than crooked southern creeks.

Meanwhile, the miles and years took a toll. The gunwales crimped and cracked, the cane seats and wooden thwarts rotted, the hull splintered on rocks and logs and suffered a long gouge from a submerged steel fence post. Scott, who's a professional boat craftsman, patched it periodically for me, and my brother Robert put in new seats when he borrowed it. I'm no craftsman, but even I replaced the wooden crosspieces in the ends when the old ones broke.

Eventually, though, I wanted a less specialized boat, not to mention a newer one, and I found one in a store in my area. The Buffalo-brand model is made of a nigh-indestructible state-of-the-art foam-core plastic called Royalex, and was in my price range, which is to say three figures. At 16 feet it fit handily in the back of my long-wheel-base pickup truck. Its shape allowed it to turn easily on crooked creeks and bayous. But it's chunky and sluggish. The Old Town routinely makes 3 miles per hour, the Buffalo closer to 2.

For the Pascagoula trip, the Buffalo would fare better on the shallow, rocky upper reaches, but would be slow on the long, wide-open reaches farther down. The Old Town, on the other hand, would be a handicap on the creek-like parts but in its element when the river widened. Which to take?

Travis faced a similar dilemma. He had a 17-foot Grumman aluminum canoe older than himself, plus a brand-new 13-foot plastic Mohawk solo model he'd purchased before learning of our trip. In the old days, when cowboys or fur trappers got ready for a long trip, they'd round up their best horses and select the ones most suited for the journey. Likewise, Travis and I brought these four canoes to a local state park one Saturday in February, two months before departure. The idea was to see which were the most efficient and compatible.

We launched at one side of the lake and stroked directly across to the other and back—a distance of just under 2 miles—timing ourselves and measuring our speed on Travis's GPS. After trying different combinations of vessels, we ruled out the Grumman and the Buffalo as too slow for the kind of pace we needed to set. The Old Town and Mohawk, on the other hand, were not only swift but moved at pretty much the same speed, which surprised me considering the difference in their lengths. Usually, the longer the boat, the faster it is, so I expected my 17-footer to far outstrip Travis's 13-footer. His Mohawk's 29-inch width made the difference. Most canoe widths are in the 36-inch range. The Mohawk was so narrow that Travis could use a double-blade kayak paddle without overreaching. The sheer mass of my canoe also made it harder to propel, slowing me down.

Speed wasn't the only factor we had to consider. Due to the length of the trip, we knew we would have to take plenty of food, gear, and drinking water. The big Old Town would easily hold all my stuff and some of Travis's, if need be. (In fact, it could even carry both of us and our gear, but Travis wanted to paddle his own boat solo, and that was OK with me.) The differences between Travis's compact new boat and my bigger, older one mirrored the contrast between us. Travis was 29 years old, 5 feet 9 inches tall, and weighed 155 pounds. I was 48, 6-foot-2, and 190 pounds. I hoped we would prove to be as compatible as our canoes.

Some rapids are too shallow for a loaded canoe to get over.

2

Though Eddie McCalip couldn't paddle with us because of his shoulder injury, he helped in other ways. Towing a borrowed canoe trailer behind his green Jeep Cherokee, he drove Scott, Travis, and me, with our boats and gear, to the town of Taylorsville in south-central Mississippi. The plan was for Eddie to drop Travis and me off at the Leaf River, then take Scott on over to the Chickasawhay. Eddie also planned to film us for the TV show *Listen to the Eagle*. Plus, Walt Grayson, a well-known Mississippi TV newsman, was interested in filming, so he and his wife, Jo, met us in Taylorsville.

Such publicity was part of an increasing trend for the Pascagoula, which was finally receiving much-deserved acclaim. A high point came in the fall of 2003 when Mississippi Public Broadcasting and the Nature Conservancy unveiled a one-hour television documentary, *The Singing River: Rhythms of Nature*. Scott and I got to attend a premiere at the Mississippi Museum of Natural Science.

The Emmy-winning film began with an aerial view of a jungle-looking forest stretching to the horizon. Mist rose through the giant trees, and a wide brown river snaked among them. It could have been the Amazon rain forest, southeast Asia, equatorial Africa. I swelled with pride at the knowledge that this was Mississippi's own Pascagoula basin.

This wild look wasn't accidental. In the 1970s Graham Wisner of the Pascagoula Hardwood Company wanted to save remaining unspoiled tracts of river-bottom timber, so he helped cut a deal between the company, the Nature Conservancy, and the Mississippi Game and Fish Commission to sell the property to the state. The deal was a groundbreaking environmental maneuver that served as an example to other states. Also helping protect the river was the rare gulf sturgeon, a giant fish which uses the Pascagoula system to spawn. Conservationists used the sturgeon to block a proposed dam on Bowie River, a Pascagoula tributary, in the 1990s. The film showed University of Southern Mississippi researchers netting and studying the weird, prehistoric-looking fish, then releasing them. Other creatures got attention too, from massive alligators to some of the 300 species of birds and 136 species of fish that frequent the swamp.

Although many people contributed to the documentary project, Kirk Loy was the driving force, serving as producer, director, editor, and photographer. At the reception following the premiere, Loy said the project began as a labor of love and turned into sheer adventure. He filmed everything from underwater fish scenes to swimming alligators to soaring eagles to billowing thunderheads.

We were helping to carry on that publicity campaign with our modest little trip, but as we drove back and forth in Taylorsville, unable to find the road leading north, we weren't getting off to an auspicious start. We needed to go north to intersect with Smith County Road No. 80, but so far we couldn't even get out of town—and Walt and Jo Grayson were following in their vehicle on the assumption that we knew what we were doing. I say "we," but I was the navigator, with atlas in lap. We eventually found the way out, but couldn't locate County Road 80. At my request we pulled over to a country store where I went inside, bought some Stage Planks—an old-timey ginger-and-molasses cookie topped with pink icing—and got directions from the woman behind the counter. As I walked out, Walt met me with a concerned, conspiratorial look. "I've told my wife men never stop to ask directions," he said.

"Directions? I was just getting some Stage Planks," I said.

He appeared relieved. "I'll tell her that. Stage Planks. Right."

We drove down a gravel road to a new concrete bridge over a clear, shallow, shady creek which was the upper Leaf River. Eddie and Walt filmed as Travis and I carried canoes to the water and loaded them with gear. After setting a heavy duffel bag in my boat, I turned to find Walt's massive camera right in my face. I ignored it like a true professional. I've seen enough nature documentaries to know how to conduct myself in front of a lens. I grew up watching the TV show *Wild Kingdom* with Jim, the manly assistant in a safari suit—though I didn't quite have his ramrod posture and deep voice when Walt interviewed me under the bridge.

Eddie, meanwhile, wasn't content with my being a star; he also wanted me to be a cameraman. He handed me a spare, compact video camera and took 30 seconds to show me how to use it, with the request that I spend about 5 minutes per day filming until the battery ran out. When I expressed uncertainty over how to operate the thing, he said I'd be able to figure it out. Little did he grasp the shallowness of my technological savvy. But we were in too big a rush loading for me to persist.

By 2 p.m. Travis and I had our boats packed and were ready to push off. Eddie suggested we paddle upstream, then float back down so he could get some footage from the bridge. Travis and I muscled upriver 100 yards or so, then turned around. Eddie and Scott stood atop the bridge while Walt filmed from the bank. As we drifted under, Scott yelled, "*No puede ir! Es muy peligroso!*"

That line was from a trip Scott and I took down Honduras's Patuca River in 1995, when the Miskito Indians warned us of fierce headwinds, waves, rapids, crocodiles, snakes, mosquitoes, and men who would cut our throats for 5 lempiras. "*No puede ir!*" they told us. "*Es muy peligroso!*"— "You can't go! It's very dangerous!" But we went, and by the grace of God survived. Ever since then, "*no puede ir*" has been our sardonic motto.

As Travis and I passed under the bridge, I knew the cameras were running, so I didn't look back, wave, or make any sarcastic remarks. We just paddled side by side, Travis in his short green boat, me in my long red one. Their pointed bows split the clear sheet of river, and each dip of our paddles left spinning, widening circles behind us. After all the commotion—loading, driving, talking, searching, unloading, filming—the embrace of silence was welcoming. Yet it wasn't really silence, because the air was full of the sounds of running water and birdsong. When we were almost out of sight, somebody yelled goodbye, and I turned around and waved.

The shallow, slow-moving river was clear but with a caramel tint, characteristic of the "blackwater" found in parts of the Deep South. The tint, which ranges from honey to cola depending on the stream, is caused by naturally occurring vegetation, rather like how tea is made of water stained by tea leaves. I've never been able to determine why some streams are blackwater and others are not. Plenty of deep-woods rivers with abundant vegetation lack the coloring, so I assume there are certain plant species or soil types that promote it. The Leaf's pale sandy bottom highlighted the tea-like coloring. The bed alternated between sand and clay, and in some cases clay-rock—not hard enough yet to be true rock, but firmer than plain clay. The water was unusually low for springtime, which in the Deep South tends to be a season of rain and flood. Current conditions were more like late summer or fall. You could see the water line along the sides 2 or 3 feet higher than the river, like a ring around a bathtub. Branches lodged in treetops were relics of real floods, when the river raged out of its banks. Now it couldn't have been gentler.

A few sections of cutover, or logged-over areas, lined the banks, but mostly there were big woods with typical riverside species: river birch, water oak, sycamore, willow, red and silver maple, baldcypress, black gum, hickory, beech, spruce pine. This being spring, the leaves looked young, tender and bright. The sycamore leaves had a frosty tint; the oaks were bright as lime slush. Cypress needles looked fluffy as cat's fur. Here and there amid all this greenery, blooming trees and shrubs stood out: white dogwoods and fringe trees, pink native azaleas. Dogwoods flourish more on piney hills than river banks, but a few make their way down to these lowlands. Fringe trees, also known as Grancy graybeard, are sporadic: I've only seen them growing wild on a couple of streams, and Travis, who knows his trees, had never seen one. They're popular ornamentals with a subtle, lovely scent exuding from the soft, fringe-like white flowers; I have one in my orchard just for aesthetics. They're in the olive family and there are only two species, the other found in China. Native or wild azaleas, also called wild honeysuckle, are different from the more prolific white and yellow Japanese honeysuckle vine. In the rhododendron family, the native azalea is a free-standing shrub that flourishes on river and swamp edges, glowing pale pink with a mild, sweet fragrance.

Travis paddled in silence. While I can appreciate conversation on a float, I value a partner who can hold his tongue and let nature do the talking. He also shared my tastes in river attire. Travis wore khaki pants, a long-sleeved blue shirt, neoprene river shoes, and a thick hemp-straw hat that was so battered it looked like he'd filched it from Robinson Crusoe. I wore olive pants, a brown T-shirt with a canoe scene on the front, a long-sleeved faded green shirt, wide-brimmed khaki hat, and rubber boots. Most paddlers these days go for shorts and river sandals, but I don't care for sunburn, mosquito bites, or wet feet.

Our trance state gave way to a pleasant surge of adrenaline as we heard the first rapid up ahead. Mississippi has no serious whitewater, but there is one stream, Okatoma Creek, with numerous Class I–II rapids, and there are many other swift stretches scattered on rivers throughout the state. The one coming up looked typical: the entire river plunged over a 1-foot drop. I stood up in my canoe to get a look. Travis's nimble 13-foot canoe should easily handle it, but I wasn't so sure about my Old Town, which maneuvers about as deftly as a battleship. Fortunately I've piloted it down southern streams enough to know how to muscle it into position. If the water were a foot deeper we could scoot over anywhere, but with it low like this we had to find a way through. A tongue of water slipping through the rocks indicated the route we should take. We bounced through with no mishap, though our pulses jumped up a notch.

I'm no whitewater expert, but I've learned on these rivers that it's usually better to slow down when approaching a rapid rather than speed up, akin to touching the brakes as you approach a yellow light rather than mashing down on the accelerator. I remember watching a group of students come through the biggest rapid on Okatoma Creek, and the adult group leader stood on the rocks yelling "Go-go-go!" at precisely the point they should have been going slow-slow-slow. Several of them capsized, which I suppose added to their fun but seemed awfully unnecessary and possibly dangerous. By backpaddling or beaver-tailing your paddle in the water, you can make your canoe go slower than the current and hence maintain more control over the vessel.

Small rapids alternated with long, smooth stretches for the rest of the afternoon. Moths swarmed over the swifts, almost bouncing off us as we floated through. One place had a double drop: we slid through the first slot and bounced over a second. Another shoal was too shallow to go over, and we had to get out and walk the canoes over. In a third, a log blocked the one way through, so again we had to wade.

The slow stretches held our interest as well. Schools of fish darted under our boats. Travis called my attention to a snake stretched out on a limb a foot from where I had paddled unawares. Another snake swimming the river paused to raise its head and eye my canoe, which must have looked like the *Titanic.*

I smelled tobacco smoke and peered ahead to see a family on a sandbar. A bald, tan, 50-something man in jeans and T-shirt puffed a cigarette as he fished with a rod and reel. A woman was sunbathing on a reclining lawn chair, the straps to her bikini top pulled down across her upper arms. When she heard us she sat up and looked around without pulling the straps up. She was thin, maybe 50, with orange-brown hair and missing teeth. Beyond them, three children splashed in the water.

"Y'all going to Taylorsville?" the man asked as we drifted by, naming the next takeout a few miles downstream.

"Gautier," I replied, citing our destination on the coast.

"Goodness!" the woman said, admiringly, I thought.

We came alongside the two girls and a boy, who played gleefully in what had to be icy water now in early April.

"Isn't that cold?" I asked them. "Or are you used to it?"

"We're used to it," a girl piped.

Farther on, another sign of people was less welcoming: wooden steps led up the bank to a camphouse with a sign reading "No trespassing. Not responsible for death or injury of anyone caught trespassing." The house was shuttered, and wire fence ringed the property. The inhospitable message contrasted jarringly with pretty, pink azaleas—domestic, not wild—that suggested a loving touch. I imagined some lunatic with an automatic pistol laboring under the delusion that his sign would protect him from criminal liability if he blew somebody away. More likely the landowners had been burglarized before and overreacted a bit with the sign and fence.

Before its batteries played out, Travis's GPS indicated we were making about 2½ miles per hour, a reasonable speed. Our goal was 15 miles a day, and at that rate we should have no trouble. Around 4:30 p.m. we stopped to check out a sandbar, but it had two marks against it as a campsite: hot afternoon sunlight and four-wheeler tracks. The next sandbar looked perfect: shady and backed by a huge cutover with no access and no human signs. We began unloading.

We had enough equipment to outfit a fur-trapping expedition to the Rocky Mountains. Travis and I were both used to shorter trips, so even we were surprised by the amount of stuff we had. But try taking food, shelter, clothing, and everything else you'd need for two weeks and see how big a pile it makes.

My largest bag was a huge canvas duffel, well-worn and faded, that zipped open lengthwise. In this I kept my tent, tarpaulin, machete, candle, propane stove, fuel bottles, grill, trowel, hammock, Captain Morgan spiced rum, and a nylon stuff bag containing a cook pot, coffee pot, double skillet, plastic plate, metal bowl, fork, spoon, can opener, spatula, and white enamel Coleman mug with the emblem of a lantern and the words "Sunshine of the Night." A large red waterproof river bag with shoulder straps held my clothes, sleeping bag, sleeping pad, and toiletries. A smaller canvas duffel carried food for breakfast, lunch, and supper: instant oatmeal, Grape-Nuts cereal, powdered milk, Luzianne tea bags, bread, peanut butter, jelly, sardines, crackers, Spanish and black olives, carrots, granola bars, rice, tuna packs, vermicelli, tomato paste, tomato sauce, chili peppers, Tabasco, creole seasoning, salt, cookies, and hot chocolate. A rubberized white bag held the items I might need across the day: a knapsack with poncho, toilet paper, lighters, Polar Pure iodine water purification crystals, compass, sunglasses, sunscreen, and insect repellent; a stuff sack full of snacks like dried fruit, nuts, and chocolate; and plastic bags containing reporter's notebooks and field identification guides for trees, birds, wildflowers, and animal tracks. A small waterproof bag held camera, film, wallet, and keys. My canoe also contained half a dozen Clorox jugs full of drinking water. The white plastic bottles are leak-proof and more durable than most canteens. I just wash them with dish detergent after the bleach is gone. For pure luxury I brought a lightweight backpacker's guitar (given to me by canoeing buddy and Mississippi newspaper owner/publisher Wyatt Emmerich on a Black Creek float) in a black nylon case packed in a green army duffel bag. My boat also held one of Scott's food bags—which brings me to a topic I've been reluctant to discuss because it might in some way suggest I am criticizing Scott, which of course I would never do. Namely, the use of a sea kayak.

Sea kayaks are admirable vessels. Eskimos invented them millennia ago out of sealskin for pursuing polar bears and walruses in Arctic seas. Nowadays they're made of plastic, fiberglass, or wood, although the wood is coated with space-age epoxy, so don't be fooled into thinking it's somehow "traditional," as Scott is sure to claim. There's nothing wrong with sea kayaks, other than the fact that they're uncomfortable, limited in the amount of gear they can carry, and impractical on small streams since they don't turn easily. The only question that arises is why anyone would bring an Eskimo hunting vessel to float a Deep South river, unless southern customs aren't good enough for them maybe. Canoes, by contrast, are quintessentially southern, having descended from the dugouts used by Native Americans of yore. Nothing could be more natural or at home on these waters than a beautiful, graceful, comfortable, spacious, streamlined canoe. Nor could anything be more practical. A canoe handles well in virtually any condition and holds all the gear you could possibly want. Scott, on the other hand, couldn't begin to pack everything he needed into his skinny sea kayak, so he asked me to carry one of his bags. I graciously agreed and never dreamed of complaining about the way it took up room and was a constant bother for me to look after.

As Travis pitched his tent, I slung my blue nylon hammock between two trees, grabbed my little camping pillow, poured a shot of rum with Clorox water into my enamel mug, and nestled down. Worn out from this long, busy day, I closed my eyes to the alternating sounds of breeze-rustled trees and humming mosquitoes. Yes.

After a while I rallied and began rounding up deadwood for a campfire while Travis cooked supper. The plan was we'd alternate suppers and provide our own breakfasts and lunches. He cooked a pot of ramen noodles with chicken. Not bad. I washed the dishes in the creek, scrubbing them with sand and feeding the remains to the minnows. Then I dragged my heavy canoe far up the sandbar, explaining to Travis that on more than one occasion I had awakened to find it floating downstream because of an unexpected rise in the river. "Reason tells me the water will never rise this high," I said, referring to the current dry conditions. "Experience tells me not to listen to reason." He pulled his boat up as well.

As darkness fell we built a campfire and sat around talking. Travis described his profession and explained his belief in sustainable forestry and low-impact logging. "That's where I get my satisfaction," he said. "Most landowners have no idea the amount of trouble we go to to leave that little oak or that dogwood tree or not make ruts." One technique is to refrain from over-thinning, or cutting trees unnecessarily. "If you don't take the best the first time around, you leave something good to come back for," Travis said with a characteristic blend of idealism and pragmatism. "Reduce cutting trees that you don't have to cut. We try to work around things rather than just knock them down and get them out of the way; not rut up the woods, and consider erosion, and things like that."

Travis had worked in the woods since age 13 or 14. His father, Doug, logs with him. "Everybody in his [Doug's] family did it at one time or another. His dad and my great-great-grandfather had a lumber mill near our house, a sawmill," Travis said. "I do the chain saws and short logs, 16 and 21 [feet]. We carry the tops for paper wood. We do it sustainably and very carefully. I do it with aesthetics in mind and the environment in mind . . . I don't cut hardwoods. I don't believe in it." He, his father, and saw hand Lafayette Brown use a John Deere tractor instead of an earth-churning skidder.

Travis, a slender fellow with dark blond hair and a brown beard, attended community college but never graduated, preferring to stick with the family logging business. Nevertheless, he's well-versed in literature, music, and art. He'd brought a copy of John Steinbeck's *Sweet Thursday* to read, Steinbeck being one of his favorite authors, along with Twain, Hemingway, Conrad, Melville, and Gabriel García Márquez. Though he played drums for a rock band, "I like the piano works, primarily—Brahms and Schubert," he said, adding that he doesn't care for Schubert's symphonic works, which he considers flowery. "I'm picky about my opera, but I do like some Wagner." Art? Homer, Turner, and Schiele are among his picks.

We turned in, Travis to his little two-man Walrus Armadillo tent, I to my four-man Eureka Timberline Outfitter. Manufacturers exaggerate tent capacities, so in reality his was comfortable for one and mine for two, or one if I spread out. I slept fitfully, as I always do the first night of a camping trip, and awakened when rain dashed against the tent around 3 a.m.

Outflow pipe from Taylorsville municipal sewage lagoon discharges treated water into the Leaf.

3

I climbed out of my tent into a drizzly dawn and quickly pitched a tarpaulin. Canoe-campers learn early on about the value of a tarp. Without one, options in the rain consist of standing around in a poncho or huddling inside a tent. I attached one end of the lightweight green nylon shelter to a log buried in a sandbank and the other to an aluminum pole, staking out the sides. Travis and I crouched underneath and made breakfast on our camp stoves. For me that meant instant oatmeal and hot tea. Travis made his trademark espresso coffee, its rich scent spicing the misty chill air. He buys Community (a popular regional brand based in Baton Rouge, Louisiana) dark-roast beans and grinds them at home. When camping or on a logging job, he uses a little one-cup espresso pot that spews the brew through a hooked pipe straight into his cup. Much as I like the flavor of coffee—I used to be a coffee drinker, strong coffee at that—I prefer the gentleness of tea these days, so I was content to smell the aroma.

The light rain subsided and we broke camp. It's no fun to roll up tents when they're soaking wet and glippy with sand, but we had no choice. I bailed the water out of my canoe, using the top half of a Clorox jug and a big sponge. Travis finished packing while I wrote in my journal, and he paddled across the river to try to identify a bush covered in white flowers, using my field guide. I finished up and decided to have my own little sunrise religious service, this being Easter Sunday. I sat on the end of my canoe and opened my little testament to the 23rd Psalm. The famous passage had popped into my head a few days before the trip. As a preacher's son I had memorized it as a child, but it grew fuzzy over the years. This time, though, it returned intact in my mind. I read silently, marveling at how well it captured the mood of a river float:

The Lord is my shepherd, I shall not be in want.
He makes me lie down in green pastures,

he leads me beside quiet waters,
he restores my soul.
He guides me in paths of righteousness for his name's sake.
Even though I walk
through the valley of the shadow of death,
I will fear no evil,
for you are with me;
your rod and your staff,
they comfort me.
You prepare a table before me
in the presence of my enemies.
You anoint my head with oil;
my cup overflows.
Surely goodness and love will follow me
all the days of my life,
and I will dwell in the house of the Lord
forever.

I looked up at the tall trees, said a quiet prayer, and absorbed the peace of the forest—a good way to start the day. Travis had drifted downstream, so I slid my canoe into the water and joined him. "What was the flower?" I asked.

"The nearest I can tell is a blackhaw," he said.

It was probably a rusty blackhaw viburnum. The bush is more like a small tree, and its white flowers grow in clusters 4 to 6 inches across. The preferred habitat is moist, rich alluvium.

We paddled along in the cool, damp morning, passing a long stretch of cutover on the left bank. The effects of the logging were obvious. The bank, no longer held together by tree roots, had caved off, filling that side of the river with sand and dirt. We had to skirt around the shoal even in our shallow-draft canoes. Travis pointed out that logging to the bank almost always produces shoals. Best Management Practices, a set of voluntary standards developed by the Mississippi Forestry Association, call for leaving a buffer zone of trees up to 60 feet wide along a river bank, but there was none of that here. Travis pointed out that a flood might have washed away the buffer zone, and suggested BMPs should call for larger swaths.

Though an environmentalist, Travis stuck up for the timber industry. All professional loggers are required to attend courses explaining BMPs and sustainable forestry techniques, he said. All timber companies have adopted the practices. But occasionally landowners—who have the final say on what happens on their property—insist on disregarding the practices, which are not mandated by law. As a result, there are too many cutovers clear to the river bank. The result, among other problems, is erosion, siltation, and elevated water temperatures, which are harmful to fish.

Just past the cutover we saw a pasture, which produces similar effects when it lacks a buffer of trees. According to the Mississippi Department of Environmental Quality, 21 percent of land in the Pascagoula basin is in agriculture, most of it pasture for livestock.

Next we saw another possible insult to the river: a 6-inch diameter metal pipe pouring liquid into the river. While Travis continued downstream, I pulled over and set out on foot to examine the pipe. The liquid splashed onto the sand and ran in a stream to the Leaf. It was clearer than the river itself, and little fish were swimming in it, a good sign. I climbed a steep, weedy bank and followed the pipeline over hill and dale, surprising a turtle with bright orange face markings basking in a mud puddle. Sewer manholes, high levees, and barbed-wire-topped hurricane fence indicated this was the Taylorsville municipal sewage lagoon.

Taylorsville is one of 37 Pascagoula basin towns with centralized wastewater treatment plants, according to DEQ. In a typical lagoon system, sewage is pumped into a pond which is deep enough to minimize odor. Bacteria in the waste begins the decomposition process. Sunlight on algae produces oxygen used by bacteria to feed on sewage. Wastewater is strained into more ponds, where it continues to break down. Then it goes out into a creek. Some towns use additional methods like aerators, chlorinators, or ultraviolet light to kill bacteria. The idea behind treatment plants is to keep the river water safe for people to swim or fish without worrying about infection, but that goal often remains elusive. Aside from towns whose lagoons don't meet safety standards, there are 100 or so unincorporated communities in the basin with no centralized sewage treatment system at all, relying instead on measures like septic tanks for individual residences. These tanks often don't work properly. Not only bacteria but pollutants like ammonia, nitrogen, and phosphorus can get into the river system as a result.

Livestock operations add to the problem with manure and farm chemicals. The DEQ's "Pascagoula River Basin Status Report 2001" estimates well over half a million cows, pigs, and sheep in the basin, and more than half a billion chickens. Factor in 143 industrial dischargers, commercial and private operations such as schools, trailer parks, and subdivisions, and 232 hazardous waste sites, and these clear, scenic streams are a far cry from what they were when Choctaw Indians paddled dugout canoes on them.

Back at the river, I pulled Eddie's video camera out of my waterproof bag. I found the on button but couldn't see anything on the view-finder screen. I tried another button and suddenly I was watching footage of some sort of outdoor festival. I hit rewind, then play, and now I was looking at a wedding. After about 10 minutes of fruitless maneuvers, I shoved the camera back in my bag and set off down the river in a huff, disgusted with technology. I like simplicity. That's why I prefer a canoe to, say, a sailboat or motorboat. I don't even carry a GPS, just a compass, and don't often use it. I garden without motors or chemicals. Until recently I used a 1965-model single-lens reflex 35-millimeter camera without so much as a light meter. I'm hopelessly antiquated.

When I caught up with Travis, he told me he'd heard a turkey gobble, but this was wild turkey hunting season and the sound was suspect. "I think it may be a hunter. It sounded a little too stiff," he said.

A great blue heron stood motionless, an osprey soared, and a barred owl perched on a tree limb. I feel a special affinity with the lanky, mind-their-own business herons, and consider them my spirit animal. The barred owl is rarely seen but often heard. Its baseline cry is the familiar "Who cooks for you, who cooks for you all," but when it gets cranked up, the variations can get maniacal. The majestic fish-eating osprey is larger than a hawk but smaller than an eagle, with dark wings, white breast, and white head with a bandit's mask. Travis claimed to have seen one the previous day but I registered skepticism since the guidebook indicates their range in Mississippi doesn't extend this far inland from the coast. Now that I saw one with my own eyes I allowed he might have been right.

As we passed the Highway 28 bridge, I expressed surprise that there was no canoe rental business, since this was such a fine river for paddling. Mississippi just doesn't have much of a canoeing culture or heritage. Unlike in northern states and Canada where the birch-bark canoe developed,

Mississippi's log dugouts evolved into plank pirogues and johnboats, and canoeing pretty much fell by the wayside. Though it's popular on a handful of streams—notably Okatoma Creek, Black Creek, and Bogue Chitto River—it's still alien to many Mississippians.

An inactive oil well stood on the left bank on a high, sandy bluff. I climbed up and examined the oily, rusty machinery. Fresh tire tracks leading down a red gravel road into the pines suggested the well wasn't abandoned. Under the riverbank, sand was crumbling around a plank-and-concrete retaining wall. I wondered how many floods it would take to collapse the site into the river. Maybe the operator would shore it up by then.

I rarely see oil wells right on riverbanks, but there are plenty away from them. The Pascagoula basin has some 2,000 oil and gas wells that produce 75 percent of the state's oil and 50 percent of its natural gas. Pipelines cross the river at several locations, and occasionally they break, as one did near a Leaf River tributary in 1999. The Genesis Crude Oil pipe dumped 300,000 gallons of crude, which flowed 6 or 7 miles to the Leaf near Taylorsville and another 27 miles downriver, requiring a massive cleanup effort.

When Eddie McCalip heard about the spill, he drove over on his motorcycle to take pictures. "They were spreading out the boom, which catches the oil, south of the Highway 84 bridge," he told me. "I started asking questions about it and they mentioned that the spill was in the river north of the bridge and they had trapped most of the oil south of the bridge and up in that direction, and they had some more places below 84 where they had the booms out. It's sort of like you see on TV. Everybody remembers the images they saw from *Exxon Valdez*. I didn't see any wildlife . . . but you could tell there was a good foot and a half to 2 feet of oil stain on both sides of the river, both banks. You could tell the oil was there. I could smell it before I ever got there. It was like driving up into a gas station where people change their oil." Now, 4½ years after the spill, Travis and I saw no sign of it.

We stopped under some clay banks to fill water jugs. The only water I consider safe in the woods is that which I can capture straight out of a vertical spring. Even without other pollution, health officials say any standing water on the North American continent is subject to being tainted with giardia, a parasite that thrives in mammal feces and can make a person mighty sick. The most pristine Alaskan stream, being inhabited by beavers and other creatures, isn't immune. A friend of mine apparently got it from drinking from a crystalline pool in the Rockies and suffered intestinal problems for months. This clay-bank spring water may not really be safe, for all I know, but I use it anyway since it's hard to imagine how it could be contaminated. Fortunately there were plenty such places along the Leaf. Water poured out of the moss-capped cliffs as if from tiny faucets, pattering onto a rainbow of clays: pink, orange, yellow, rust, purple, brown, gray. We held our jugs under the spigots, which just trickled, so it took a long time to fill. The water was cool and woodsy-tasting.

At lunch on a sandbar we ate peanut butter and jelly sandwiches, our meal observed by a turtle on a log with a pale, dirty back. Four yellow and black butterflies, tiger swallowtails, dined at white bird droppings in the mud. Songbirds twittered from the woods on 15-to-20-foot-high banks behind us.

The afternoon stretched before us with extraordinary peace as we resumed paddling. A wild turkey gobbled three times in two-minute intervals, and this one sounded real. An osprey glided overhead, whistling. Black vultures, a dozen or more, clustered in treetops with the shifty demeanor of hoodlums; I half-expected to see cigarettes dangling from their beaks. A red-shouldered hawk burst out of the woods beside a power line shrieking; it was identifiable by its rust-colored breast, unlike its white-chested cousin, the red-tailed hawk.

Oakohay Creek flowed in from the right, wide, shallow, and tea-colored beneath big trees. In 1990 Scott and I launched our canoes up this creek near the community of Hot Coffee and paddled down to the Leaf for a week-long float. Now Travis and I landed on a sandy point at the mouth and he made coffee while I walked up a sandbar layered with cracked mud and pocked with deer tracks. The Oakohay—some call it 'Kohay for short—was clearer than the Leaf, and logs across the stream suggested it would be difficult to negotiate in low water.

At the four-lane Highway 84 bridge, I stopped at a boat ramp to check out the reception on my cell phone. I had never taken cell phones, radios, or other such devices on river trips, and I've always made fun of people who do. But this time my mother, Maude Riales of Memphis, Tennessee, was sick and I needed to keep up with her condition. *Listen to the Eagle* host Paul Ott loaned me a cell phone from one of his sponsors, and in return I agreed to call his show each Monday night. It felt strange carrying a phone on a canoe trip, but I had to admit it was a smart idea. Not only would I get to hear my wife Angelyn's cheerful voice every two or three days, but if Travis and I had problems—such as snakebite or injury—a phone would make things much easier. Hearing Angelyn on the phone just now was a spirit lifter. While we chatted, traffic roared past on the highway, and a teenaged girl rode a four-wheeler back and forth under the bridge, stirring up dust, as her parents watched.

At the next sandbar down we passed an overweight woman and boy and an underweight man with a malamute-looking dog. "I just got me a husband," the woman bragged to us. The guy, dressed in cutoff jeans, had stringy, graying black hair, a goatee, and tattoos.

"Beautiful dog. Is that a malamute?" I asked, admiring her pet.

"Malamute-wolf hybrid," she said.

I told her about a man I know who raises wolf-hybrid pups, and then the current carried us out of earshot.

We stopped to check out a low-lying sandbar for camping. An ideal campsite lacks easy access to humans from the land and has firewood, trees for hammocks, and shade. We found no tire tracks but also no firewood or hammock trees. Plus a great blue heron—my spirit animal—swooped out of a tree and headed downriver as if to lead us on.

The next sandbar looked good except for some old tire tracks. But we were tired and Travis argued that since it was Easter Sunday evening, no one was likely to show up. We were wrong. As we pitched our tents, a pickup truck pulled out on the sandbar and a man and two boys got out and walked toward us.

The law is not crystal clear as to whether it's legal to camp on sandbars without landowners' permission, but I'd hate to test it in court. My philosophy is to choose the most remote sites I can find and leave nothing but footprints and maybe campfire ashes. In three decades of paddling I'd never had a run-in with a landowner—not even an encounter that I recall. My attitude is that if someone asks me to leave, I'll do so.

As the man approached, I noticed he had a gentleman-farmer look: straw derby hat, white hair, spectacles, pale blue shirt buttoned to the throat and tucked into pleated khaki pants, brown loafers. Travis and I introduced ourselves and shook his hand as he launched into a string of polite but unrelenting questions. Where did we launch? Where were we taking out?

Why were we making the float? Where were we from? What did we do for a living?

After he was satisfied that we weren't a pair of thugs, he introduced himself—to our amazement—as Charles Pickering. He didn't say "Federal Judge Charles Pickering." He didn't have to. He'd been all over the news recently, including an appearance on *60 Minutes*. President Bush had used a recess appointment to name Pickering to the United States Court of Appeals for the Fifth Circuit, prompting outrage from Democrats. The worst critics called Pickering a racist. Many people in Mississippi, including some black civil rights leaders, said he was not, and that he was guilty of nothing more than being a white Republican male from Mississippi. Travis had seen the *60 Minutes* appearance and chatted with the judge about it while I wondered just how deep in the jailhouse he could put us for trespassing on his property. But, having learned where we were from, Pickering quickly established mutual acquaintances: retired barber Clifton Terrell of Liberty, retired Senate sergeant-at-arms Charles Ravencraft, deceased county Republican leader Dewitt Smylie. Then he asked to see our boats.

It was obvious that we were off the hook. Judge Pickering seemed impressed, even enthusiastic about our journey. He only hinted at the trespassing issue by asking if we'd checked land ownership along the river to decide where to camp. Considering the fact that our trip spanned 210 river miles across eight counties (Smith, Covington, Jones, Forrest, Perry, Greene, George, and Jackson), the answer was self-evident. Besides, he acknowledged that he had done a bit of canoe-camping himself over the years and stayed on sandbars just like we were doing.

Judge Pickering owns much land along the Leaf and grew up near it. "One of my earliest and fondest memories is when we would finish laying the crop by in the summer, I would go with my dad and neighbor and they would set trotlines," he said. "One summer—I suppose I was probably 8 to 10—they knew where a hollow log was; it was right at the mouth of Mason Creek. After we'd set the trotlines, they went over to try to retrieve a catfish out of that hollow log, and eventually did retrieve a 48-pound catfish out of that hollow log."

Pickering was describing the practice of fish grabbing, also known as grabbling, hand grabbing, and noodling. From late spring through early summer, catfish make dens in hollow logs and holes in the bank. The female lays eggs and the male stands guard. Grabbers locate the fish with thin sticks, plug escape routes with gunnysacks, then reach in and pull the fish out, either bare-handed or with grappling hooks.

Pickering has also hunted along the Leaf for deer, turkeys, raccoons, wild hogs, and squirrels. "I've carried Chip squirrel hunting along Mason Creek and Leaf River," he said, referring to his son who was now a U.S. Congressman. "We used to hunt squirrels with a dog, a little feist dog. We've eaten Vienna sausage and crackers on the Leaf River banks while we were resting during the middle of the day."

In the early 1990s, wild hogs started turning up on his land. They compete with native animals for food and also root up the banks. Landowners hunt them with dogs or catch them in box traps baited with corn. "The first hogs that I had in that area were around 1992, and we caught, trapped, or killed 48 hogs," Pickering said. "We got them all except one, and then a neighbor killed that hog a few years later . . . We had no hogs for 10 or 12 years, and they've just come back within the last year." Pickering had trapped 8 so far this year and was interested to learn Travis and I had seen fresh hog rootings while we were checking out his sandbar.

Pickering has seen other changes along the Leaf over the years, including falling water levels. "The Mason Creek was a good flowing stream that you could fish in and it later largely dried up. It has a little more water in it now than it did for a few years," he said.

"The land became more open as more land was cleared up, more timber was cut. You're seeing some reversal of that and seeing more timber planted now, but that had an impact on it. I think it probably is not as deep as it was back in the '50s. It was deeper when I was growing up. I don't know what the cause of that is.

"There's more boating on the river than there was, although there's still not an awful lot. And you see some intrusion of four-wheelers. We have had some problem with four-wheelers and have had some problems with the use of drugs, camping, and just people under the influence, being rowdy. The game wardens cut down on some of that. People are supposed to have written permission" to ride four-wheelers on others' property.

Pickering was among sponsors of legislation to buy the land to be set aside as Pascagoula River Wildlife Management Area. "When I was in the [state] senate I was involved and instrumental in the purchase of the Pascagoula Hardwoods," he said.

"You've got a tremendously large tract of land that's preserved for this generation and for future generations so that they can have a wilderness experience, and maintain the beauty of it," he said. "I think it's very important that we have those kinds of areas that are available for the public, particularly as more and more private land is not accessible to the public."

Travis and I, who would be canoeing right through the wildlife area, were direct beneficiaries of the action. I snapped a picture of the judge and his grandsons, 15-year-old Jeremy Dunkerton and his 4-year-old brother, Thomas. Pickering bid us enjoy our night on his sandbar.

Easley pilots canoe up cypress slough adjoining the Leaf.

4

Rain fell for a long time after we went to bed, then stopped way up in the night. At dawn I thought I heard footsteps outside the tent. Then I realized it was puffs of wind. A cold front was arriving. When I got up, my little camp thermometer showed 58 degrees. Travis built a fire, and I toasted bread over it, then spread the toast with raspberry preserves to go with oatmeal and tea.

Setting off down the river, we saw an unusual-looking white heron with black markings on its wings. My field guide identified it as an immature little blue heron. Both great and little blue herons start out white before taking on their adult plumage, and some egrets are also white, so sorting them out can get tricky.

I'm not a natural-born naturalist. On a field trip to the beach in ninth-grade biology class, I climbed over the sea wall and took a nap while my fellow students looked for mollusks or whatever. But I tarried a bit too long, and when the teacher deduced (or someone told her) what I'd been doing, she said there'd be no more field trips thanks to me. Despite such experiences, the biology bug never did bite me, but it's hard to spend years in the outdoors without learning to identify basic species. I'm OK on the more common trees, birds, and mammals, but offbeat critters like an immature little blue heron send me to a guidebook.

I did recognize some young buckeye trees with red blooms growing along the banks. Buckeyes are best known for their smooth, curvy nuts that old-timers kept in their pockets for lucky charms, but once you learn to recognize the trees, their leaves and flowers are pretty distinctive. I also spotted trumpet honeysuckle vines with bright orange blossoms, and thickets of white-blooming dewberry with burgundy and green leaves on their thorny ground-hugging vines. We were too early to get any berries.

Mindful of my obligation to spend a few minutes each day filming for Eddie, I stopped on a high bluff to video Travis canoeing. But I couldn't make the camera work, and I gave up in disgust. Travis, who had done some filming in the past, volunteered to take it over, to my relief. So much for my brief foray into movie production.

A pushy tailwind and clipping current sped us along through towering forest. A 1½-foot-long alligator snapping turtle lumbered past under the surface of the clear, shallow water. These turtles, often miscalled loggerheads, can get monstrous. I've seen one as big as 174 pounds that was taken live from the Mississippi River. Its face had all the charm of a single-toothed jack-o'-lantern, the claws were as long as a bear's, yet the eyes seemed surprisingly gentle—unless you're a small fish. From the tip of its tongue dangles a strip of flesh which looks and moves like a worm. I'm told the turtle lies underwater and uses the tongue for a lure, gulping the fish that go for it.

Water rushed through a grid of downed logs up ahead, and I sized up the route to take. I noticed an old bridge piling standing by itself but didn't pay much attention since it was off to the side. But as I angled my boat between a pair of stobs, I suddenly saw a horizontal iron bar directly across my path in the rushing current. I made a sharp left turn and skidded into the shallows to avoid collision, motioning for Travis to do the same. With the fast water and immovable bar, a crash could have been nasty. We had to drag our boats around the obstacle.

A red camphouse literally tossed up among some trees bespoke high water and bad planning. I'm forever amazed at camps, and sometimes residences, built in vulnerable places on riverbanks. Apparently the owners think a stream is like a street. But rivers over time whip and thrash violently, collapsing banks, cutting new channels, bringing heavy timber downstream in floods. I'd rather build a camp well away from the river and set a couple of chairs on the bank for the view—which many people do.

A shady slough entered from the left, and Travis and I nosed into it with our boats. Cypress trees loomed over the dark enclosure. A few mosquitoes hummed in the shadows; there would have been more if the temperature had been warmer. We didn't get far before logs blocked our way, and we returned to the river. Often such sloughs provide wildlife sightings. I've seen wild hogs, owls, otters, and beavers in such places.

Below Highway 588, an unusual sight on a small sandbar to my left grabbed my attention: a mound of sand with a hole in the side, a foil-covered block lying nearby, a wooden cross, and sprigs of greenery.

An Easter shrine! This was the day after the holiday, and the little display obviously represented the empty tomb of Jesus. The sight was refreshing. You're more likely to see beer cans, burned tires, and assorted detritus on sandbars. The image of someone enacting a reverent ceremony made for a pleasant contrast.

Sandbars vanished, replaced by long stretches of clay banks topped by moss and big woods. Springs and small waterfalls splashed into the river, making sweet music. Scents of vegetation seeped from the forest. Wild azaleas glowed like giant bowls of strawberry sherbet. Four antlerless deer at the water's edge bounded up the bank, black tails down. Then the rear one snorted and lifted its tail to reveal a white flag of warning. Travis and I stopped to explore a spring branch on foot, a delightful brook running over a clay bed between high banks and tall trees.

Poor Scott. Here we were, blissfully enjoying nature, while he toiled away on the Chickasawhay counting snags. We were on a pleasure cruise; he was a galley slave. Yet the fact that he was getting paid diminished our sympathy. He was making money for floating a river; how bad could that be?

With no sandbars available by late afternoon, we climbed a high sandy-clay shelf beside a creek. I phoned Angelyn, who said it was raining at our house 100 miles to the west. It soon did the same here. Travis and I pitched tents and I strung the tarp for him to cook under. He prepared instant mashed potatoes and dehydrated ground beef with Italian seasoning that left a rank aftertaste, a cut below the rice and lemon-pepper tuna I'd made the night before. We had unwittingly entered a competition for worse camp chef.

It was Monday evening, so I called *Listen to the Eagle* on the cell phone and chatted with Paul Ott and his daughter and cohost, Carla Tigner. I couldn't get used to making phone calls from a riverbank, especially since I used to ridicule fellow campers who did the same. "Hi, honey," I'd mimic in a mincing voice. "We're making camp now. Everything's all right, sweety-pie." The shoe was on the other foot now. "I never thought I'd be calling a radio show on a canoe trip," I told Travis sheepishly.

Far from sympathizing, he retorted, "I never thought I'd go canoeing with somebody who did." I was getting my comeuppance, all right.

I walked to the edge of the high bank and gazed out over the water while rain tapped in the dusk woods. A great blue heron swooped down the river, below my level, straight toward me, then saw me and turned away. I got out my backpacking guitar and played some blues, followed by Celtic, hymns, and bluegrass.

It turned cold in the night. A critter rattled the cookpots, waking me—raccoon, opossum, skunk?—and I shouted at it: "Get out of here!"

Dawn was 43 degrees, damp and breezy. Today we were supposed to stop over at the camp of Terry Gibbs and his wife, Pat, longtime owners of Black Creek Canoe Rental at the town of Brooklyn who had recently sold the business. They owned a camp on the Leaf and had invited us to pitch our tents there for the night. We'd made such good time with the tailwind that we only had 10 miles or less to cover, a cinch considering our average speed was 2½ to 3 miles per hour.

We passed Highway 590, a lonely rural bridge, and continued through a canyon with high clay banks, shadowy woods, and few sandbars. Crashing water up ahead announced a rapid, and as we rounded a bend, it looked bigger than any we'd encountered. As usual, the entire river dropped over a rock shelf, but in this case there was a second shelf just beyond. Each drop looked to be 1½ to 2 feet. Travis went through first and took on water from the waves. He stopped in the stillness below to bail. I aimed my red battleship down the tongue of water gushing through the rocks and bounced through, ruddering to the right and then left to hit the slot on the second drop. Unlike his pint-sized boat, my ship remained dry.

Belatedly, Travis pulled out the video camera as I came through. Heroically, I volunteered to drag the boat upstream and do it again for the camera. Paddling to the far side of the river where the current was lighter, I waded in my rubber boots, pulling the canoe upstream. It was hard work, but I made it up, turned the boat, got in, and rode the rapids again. Alas, Travis was having trouble with the camera, and only got me coming through the second drop—a tragic loss for the world of adventure documentaries.

Swallows swarmed in the churning water below the falls. Usually with such birds you just glimpse a flitting silhouette, but these were all around us, as if congratulating us on our passage. I identified two species: barn swallows and bank swallows. The former had steel-blue backs, the latter brown. Barn swallows typically inhabit old buildings, as their name suggests, but their natural habitat is precisely this: rocky ledges over streams.

Bank swallows nest beside rivers, and we often saw their holes in clay banks. The birds made a chittering sound as they rushed around us in an ecstasy of feeding or just playing.

The air wasn't warming a bit. If anything it seemed to be getting chillier, and an intermittent drizzle kept me in and out of my poncho. My cold-weather gear consisted of a wool sweater and a wool shirt, and I didn't like to wear both at the same time in case they got wet. For lunch we huddled under a sandbank out of the wind and watched the gusts paint silver swirls on the water—pretty, but an unwelcome sight to a paddler. Pink native azaleas glimmered on the bank across from us. Dewberry stalks, winged elm saplings, and privet hedge grew in the sand where we sat, which was mottled with cattle tracks. Climbing the bank, we could see open pasture, an old shack, and Interstate 59 in the distance, with tiny vehicles crawling back and forth.

Terry Gibbs had told me his camp was about ¾ mile below the I-59 bridge. He said he'd mark his property with three strips of orange flagging tape. We passed under the four-lane bridge and encountered several small rapids. One was actually a fishweir, an ancient V-shaped line of rocks designed to funnel fish into a wooden trap. The Fishtrap Bluff Fishweir was on the National Register of Historic Places, though there was no marker at the site. I noticed the unusual straightness of the lines of stones but didn't realize its purpose until later.

John Connaway of the Mississippi Department of Archives and History, who has written a book on weirs and nominated this one to the national register, says some older than 5,000 years have been found in North America. "The basic design is a V-shaped dam with an opening for a trap at the downstream apex of the V," he wrote in his report on the Fishtrap Bluff site. European settlers continued to use weirs as they moved into the country, but the basic design remained. "The function of the weir also remains unchanged through the ages, that of mass procurement of fish for food at times when local communities were dependent to some extent on natural resources for subsistence," Connaway wrote. "At a certain time of year when some species of fish are spawning, these weirs were operating at their peak and were of great economic importance to the people who used them. They provided an easily acquired meat source in areas and at times when meat was not plentiful. This was especially true during prehistoric times, during the early European settlement of this country, and even during the depression era of the 1930s."

The Fishtrap Bluff weir dates to the late 1800s, Connaway determined. "It is not known exactly what year the weir was built, but according to the present owner, Bud Smith, his grandfather Henry C. Smith and some other unidentified men built it. Henry was born in 1862 or 1863, so by 1880 he would have been about 18 years old and presumably mature enough to attempt such a construction. . . . The taking of game fish by means of fishweirs was outlawed in Mississippi in 1922 (Laws of Mississippi 1922), but some continued in operation with certain restrictions (essentially that they not completely obstruct the passage of fish or any body of running water) into the 1930s. Thus it is suggested that this one was built and in operation sometime between the late 1870s and the early 1930s."

The Fishtrap Bluff weir, which Connaway numbers among the best preserved in the state, provided an important function in its day. "According to the descendants of its builder, it served to feed numerous people in the surrounding area, where the community was scattered on farms and small towns," he wrote. "Wagon-loads of fish were reportedly hauled from the weir, and people came from as far away as Mobile, Alabama, to trade for goods in the area, including fish caught there."

Cold rain fell as we zipped past, and I pulled on my poncho again. Banks as high as 100 feet rose on either side. Terry Gibbs's markers did not appear, though we passed numerous camps peeking through the leaves on the high bluffs. After we'd gone more than ¾ mile, we stopped and I got out my cell phone. Apparently I had become totally dependent on the technology I had so recently spurned. I got Terry immediately, and he said he hadn't expected us this early so he hadn't put the tape out yet or even arrived at the camp. I described where we were—at a sandbank that had collapsed beneath the front half of a camphouse under construction—and he directed us a short distance back upriver to his spot.

We followed his directions and spotted a small travel trailer atop the wooded bluff. We tethered our canoes and picked our way up the slippery bank. The Gibbses' trailer nestled among mature hardwood trees next to a pitched tarpaulin, plastic picnic table, and firepit. A gravel lane led to a neighbor's rustic camphouse 100 yards away. Travis rustled up some firewood and started a welcome blaze in the windy, 51-degree weather. At least the rain had stopped. I stretched my hammock between two trees and was settling in for a nap when Terry drove up in a pickup truck. Gray-bearded, slender, and energetic, the 60-year-old longtime canoe outfitter offered us a choice of cheap scotch or Yukon Jack as he began unloading firewood.

Anyone who canoed Mississippi's Black Creek in the 1990s and early 2000s—and that would include most paddlers in the region since Black Creek is a top destination—has probably run into Terry and Pat Gibbs. They owned the Brooklyn, Mississippi, business from 1990 to 2004 and continued to work there after they sold it. Terry was involved in sales for most of his life before he got the urge to run a canoe business. Born in Florida and raised on a farm in Alabama, he opened a mall in Jackson, managed a department store in Laurel for 20 years, and worked as a traveling sign salesman. Meanwhile, he and Pat paddled as a hobby, he told me as we passed around a clear plastic soft drink bottle of scotch that Travis compared to gasoline.

"We'd done Black Creek, Leaf, Okatoma," Terry said, ticking off Pascagoula tributaries. "We'd done Okatoma till we were blue in the face." He toyed with the idea of running a canoe rental, and in 1990 took a step which changed his life. "I believe in the Lord and I'm a Christian. I stopped off by Black Creek Canoe Rental and said, 'Have you ever thought about selling it?' and he said, 'I listed it the day before.' I took that as divine intervention." He and Pat bought the business, and though they didn't make as much money as they had been, "It's one of the best things I ever did," Terry said.

Over the years they became well acquainted with problems that beset the creek and other streams in the region. While most of Black Creek runs through the De Soto National Forest, development has increased on private property along it, notably a subdivision in the Camp Dantzler area west of the national forest. "That sewage lagoon is right on the creek," Terry said, expressing concern about what a flood could do if it overflowed the lagoon. "I don't want to do sour grapes because I have a house on the creek and those folks have a house on the creek. As long as you don't trash it and you take care of it . . . ." Ironically, bank erosion is gnawing its way toward his house, a force he wasn't familiar with when he bought it. "Every time a flood comes, it eats back and back, and it could easily undermine our houses," he said.

The Gibbses haven't lost their taste for canoeing and camping on Black Creek, Leaf River, and other waters in the Pascagoula basin. "Keeping it free-flowing—the only place in the lower 48—is what makes it attractive to me," Terry said of the basin. "No dams. A lot of access. It is still free-flowing and you can participate by the public." He referred to the national forest.

"I still get excited about great blues," he added, indicating the big herons. "It seems silly because I've seen a million of them. I've seen everything from pretty good-sized gators to all kinds of birds." Other common species continue to intrigue him as well. "We've gotten beaver-slapped many times, I mean right next to you," he said, citing the big rodents' habit of slapping the water with their tails when alarmed or irritated. "It scares Pat to death." Less common are otter sightings. "That's neat because they're elusive. You'll see that nose come up and whoops! they're gone."

The Gibbses sometimes take groups out on moonlight floats. "Just about three or four nights before your full moon is best because the moon comes up early," Terry said. "I know the creek well. I don't really have to wait for a full moon." One of the best nighttime sights is glowworms. I didn't know about these little wonders until Terry told me to watch for them on an October float. Peer into the darkness under the banks and gradually you start to see them, like a miniature galaxy.

Our talk turned to logging and its effects on rivers. Terry expressed aggravation with people who log their land clear to the streambank. "People are so stupid. You're losing your property, fool. Of course, you see some replanted. It's only a few bucks to leave some" trees, he said.

Pat arrived and started some cabbage and hamburger soup in a Crock-Pot. The mouth-watering aroma mingled with the smell of campfire smoke. Pat served the hearty soup with broccoli salad, rolls, and leftover chocolate Easter candy for dessert. You'd think Travis and I had been in the wilderness for weeks, the way we gorged. Perhaps it was the prospect of the rice, tuna, and instant potatoes that lay ahead.

Night fell and we sat contentedly around the fire, talking about outdoors topics. Then headlights appeared and a Jeep rumbled up. Terry's camping buddy, R. T. Lowe, had arrived after practice with his church bluegrass group. Bearded and jovial, he poured a mug of coffee and pulled out a guitar, singing some bluesy Leon Redbone-style numbers. He prevailed upon me to fetch my little guitar from the canoe, and we teamed up on hymns, country, and folk-rock.

So this is how Scott and I will probably be in several years, I pondered. Scott has a half-finished camphouse near Bowie Creek, and I've got a covered deck by my pond. Someday he and I will no doubt do just like Terry and R. T.: get together on occasional weekends, build a fire, have a couple of drinks, make music, talk about trips and equipment, and camp out. There are worse fates.

It was 11 p.m. when we turned in. Pat left for the house, and Terry and R. T. bedded down in winter sleeping bags on cots under the tarp, their preferred means of camping, tents being too confining for them. I, however, was happy for the snugness of my tent as I burrowed down in my sleeping bag and put in earplugs to muffle the sounds of the interstate, the one drawback to this excellent setting.

This waterfall on a Leaf tributary is one of the many surprises in the Pascagoula basin.

5

A white mist settled over the river in the night. As we loaded our canoes in the 38-degree dawn, I could barely make out the ragged green forest on the far bank. While Travis and I packed, Terry and R. T. dropped a truck off at the next bridge, Eastabuchie (East-a-BOO-chee), so Terry could join us for a morning float. Terry slid his white plastic Dagger Revelation canoe into the still Leaf River as Travis and I got into our own heavily loaded boats, their interior hulls grimy with mud and sand. Terry took a 5-gallon plastic bucket containing whatever odds and ends he might need. Fitted with a custom screw-top lid, such buckets were what he and R. T. used to carry gear, whether camping or day-floating. With decades of canoe-camping under their belts, they had their methods down pat.

I have my methods, too, and some are quite different, as became apparent when the topic of paddles came up. I think it should be self-evident that wooden paddles are the best. They're smooth, warm-feeling, and natural. They're also cheap, at least the ones I get. I use Indian Head paddles made in Ripley, Mississippi, so when I spend $16 or so for each, I'm helping support the state economy. Terry, on the other hand, had his own preference—some sort of expensive, carbon-based, bent-shafted thing, probably made in Oregon or California and designed by aerospace engineers. He explained that you use it with a J-stroke—as opposed to the guide stroke which I favor—and it's lightweight and easy to handle. Travis, meanwhile, chimed in with his own opinion, to the effect that cheap aluminum paddles are best, especially if you get the shafts encased in plastic to prevent them from turning your hands black. The three of us swapped paddles, and after a brief test I pointed out that Travis's paddle was a piece of junk, Terry commented that mine was heavy as a log, and Travis said Terry's might be good for people who could afford it.

Our debate was sidetracked by the appearance of a pair of cormorants overhead, soaring among black vultures. Terry and Travis agreed that was an odd conjunction of species. I wondered how they identified cormorants so easily considering the birds were about 500 feet up in the air. It was becoming apparent on this trip that Travis was the real naturalist between us. Even though he claimed not to know much, he tended to identify birds and plants quickly while I fumbled with my field guides. I attempted to showcase my own knowledge by pointing out the native azaleas blooming on the banks, only to have Terry inquire whether some of them might be mountain laurels. I explained that we hadn't seen any mountain laurels, though I'd been watching for them. But when he wasn't looking I took a closer peek—the bushes were way up the side of the bank—and noticed, you guessed it, a mountain laurel.

In my defense, both species are pink and showy, and you have to get pretty close to distinguish the azalea's honeysuckle-shaped flowers from the mountain laurel's multisided blooms. The book *Identification, Selection and Use of Southern Plants for Landscape Design* by Neil Odenwald and James Turner describes the latter as "round, cup-shaped with starry, crinkled edges, one inch across in large terminal clusters. Rose to nearly white, marked inside with purple flecks. Cluster of buds very attractive. Bloom March to May, depending on location." I couldn't put it better myself. The laurel grows as an understory in mature forest and doesn't fare well in cutover, and the book says it's disappearing at an alarming rate.

Soon we heard the roar of a rapid that Terry said was the biggest on the entire Leaf River. I'd come this way years before, when water levels were substantially higher, and found a river-wide 3-foot drop. Now the low water revealed a broad rocky ledge with a long, rollicking chute to the right. Wanting to get photos of my buddies, I went first, piloting my battleship right down the middle of the chute. Crashing waves and side-currents had no effect on the big red vessel. At the bottom end I stopped, climbed out on the rock ledge, and aimed my camera as my pals descended. Terry went first in his white boat and pulled aside in an eddy to wait for Travis, whose little green cockleshell bounced through, its sides too low to repel the waves. By the time he got down to me he had shipped a few gallons.

Terry led us on foot into the woods along a tributary to a broad, 6-foot-high waterfall, seemingly out of place in these Deep South lowlands. Except for the chilly weather, this could have been Hawaii. Then we floated the remaining distance to the Eastabuchie bridge, where Terry gave us some bread, chocolate, and bottles of drinking water before bidding us safe passage and heading off in his pickup truck—a good guy, despite his taste in paddles.

As he drove off, one of the lenses of my prescription sunglasses fell out of the frame, and the button popped off my pants: equipment failure, and we hadn't even been gone a week. Fortunately I was able to snap the lens back in; otherwise I would have had a hard time on the sun-glittering water for the next several days. And I cinched my belt up a notch to hold my trousers in place.

The morning was warm and sunny as we slipped past a succession of sandbars and thickets of young willow, hardwood, and pine backed by cutover. A brisk, glorious tailwind hurtled us downstream. At lunch Travis went off to bathe in the river while I wrote in my notebook and felt homesick. As a kid I was the type who called Mama to come get me after a couple of days at summer camp. In my 20s, on the other hand, I took off on backcountry trips for up to two months at a time even though I was married (to the right woman, I might add). Such journeys got progressively shorter over the years, and now I was edging back toward my childhood state, except instead of missing my mama, I missed my wife. One thing I have learned, though, is that journeys may seem long at the time but are over before you know it, never to come again. So I shrugged off my loneliness and, with a newly clean Travis leading the way, continued down the Leaf.

On this crooked river, tailwinds sometimes became headwinds, a foretaste of what awaited us. I was sitting in the stern of the canoe with the bulk of the 17-foot boat in front of me, and it proved hard to handle in those gusts. The wind tried to push the vessel sideways, requiring sheer force to right it. The old ship was creaking, too. I hoped it would survive the trip. With Travis well ahead and out of earshot, and the wind whisking my voice away, I comforted myself by singing hymns: "I Need Thee Every Hour," "Standing on the Promises," "Amazing Grace."

A mile or so above the town of Petal we came to a brand-new four-lane bridge, still unfinished. We beached and walked up onto the mammoth edifice, which gave us a panoramic view of the Leaf. From this height, its striated sandy bed was visible from bank to bank through the shallow, tea-colored water. To the east we saw traffic and train cars, signs that we were approaching the urban sprawl of Hattiesburg.

A short distance downstream, a large pipe discharged pale green water from the west bank. Looked like another sewage lagoon. Just above the

Highway 42 bridge at Petal, we turned up the mouth of Bowie River. A strong current dissuaded us from going beyond the first sandbar, where we stopped to stretch our legs. Travis made his afternoon espresso while I shook my head at the thickets of Chinese tallow trees and privet hedge, both highly invasive nonnative plants. Mitigating my dismay was the sight of fish swimming in the clear river and the sounds of birds chirping in the woods.

The 55-mile-long Bowie starts northwest of here and runs through deep woods, receiving the whitewater Okatoma Creek before passing through a stretch of gravel mines and entering the Leaf. Bowie is one of the prettiest streams in the state but is rarely floated, being upstaged by the wildly popular Okatoma with its series of rapids.

Right around here was the site of a proposed dam that would have ended the Pascagoula's status as the last major unaltered river system. In the 1990s the Pat Harrison Waterway District, the cities of Petal and Hattiesburg, and the Forrest County Board of Supervisors teamed up to fund a feasibility study for a $12 million-to-$30 million, 1,200-to-2,000-acre reservoir on Bowie. Supporters of the reservoir said southeast Mississippi needed a new water source to handle the growth of development, even though adequate groundwater was present. Opponents said a dam would disrupt the spawning habits of rare species such as the 200-pound gulf sturgeon and the tiny pearl darter and Alabama shad. The plan ultimately fizzled, to the relief of those who cherish the Pascagoula's free-flowing status.

There are in fact many dams in the basin, though they're all so far up the tributaries they don't significantly affect the Pascagoula's natural flow. These include eight flood control reservoirs that double as water parks (Archusa Creek, Big Creek, Dry Creek, Flint Creek, Little Black Creek, Maynor Creek, Okatibbee, Turkey Creek), two state parks (Paul B. Johnson and Clarkco), plus several state fishing lakes (including Lake Claude Bennett, Simpson County Lake, Lake Ross Barnett, Lakeland Park, Lake Bogue Homa, Lake Mike Conner, and Lake Perry). Factor in every pond and private lake within the basin and these man-made bodies of water tally 1,680, according to the DEQ. In that respect, nothing is "natural" anymore about the Pascagoula—or probably any other river system.

I did feel cheered when I calculated we had already come 65 miles. By the end of the day we would be a third through the journey. Two weeks may be a short time by the standards of real explorers, but to us ordinary folks it's a long time for a camping trip!

Back on the Leaf, just below the Highway 42 bridge on the left stood a massive red-brick power plant with twin smokestacks and a sign warning trespassers to stay away. Next was an iron bridge, then a red sign with white letters: "America's Holocaust: Abortion." An odd message for a river bank, but I presumed it was aimed more at motorists on the bridge than boaters.

Some folks fished from the west bank a short distance from a big iron pipe trickling liquid into the river. On the east bank, mobile homes lined a high dirt bank, down which all sorts of rubbish had been dumped in a vain attempt to stop erosion. One trailer had a Rebel flag in the window, an old Dodge pickup truck parked outside, and items such as an upside-down lawn mower and rusty bedsprings thrown down the bank. A 6-inch white plastic drainpipe protruded from the bank directly over the river, presumably either for raw sewage or "gray water" from sinks, dishwasher, and washing machine.

The air thickened with the smell of human waste, and we saw the long, high, grassy levees of the Hattiesburg sewage lagoons. My topo map showed four contiguous ponds. Murky brown outflow tumbled down a pile of gray rock riprap and filtered into the Leaf. The stench was profound. A river bend took us away from the lagoons, then back, enhancing our agony. While Travis paddled quickly on, I tied to a tree and climbed a levee, picking my way over rock rubble crowded with red-fruiting dewberry vines. From the top of the levee I gazed out over a vast, square lake dotted with what looked like spinning round bales of hay—aerators churning the water to disperse and destroy bacteria. Odor notwithstanding, it looked modern and efficient.

It was well past time to make camp, but we kept paddling to escape the smell. Around 5:30 we stopped on a sand and gravel bar a couple of miles below the lagoons. The stink was gone, but Travis found some sort of medical waste—an IV, colostomy or diabetic bag—on the beach. Our disgust was unspeakable.

We ate spaghetti to the wail of a not-so-distant train whistle. I kindled a fire and picked guitar beside it in the dwindling daylight. As night fell I pulled on my wool shirt and set off alone down the sandbar, which was at least a quarter mile long. The awfulness of the city seemed more remote now as the sand squeaked underfoot and I inhaled sweet scents of nighttime foliage and admired the far-flung stars. At the lower end of the bar, a beaver slapped the water with its tail and a deer snorted. Water sang among downed trees, suggestive of good paddling ahead. Chilly in the night air, my spirits restored, I returned to the welcoming glow of the campfire.

Dawn arrived with 41 degrees, but the rising sun heated the sandbar quickly. I built a fire, boiled water, made tea and oatmeal, grilled toast. Travis slept late for once: he said he pulled his sleeping bag over his head at 5 a.m. and never saw the sun rise. Anticipating headwinds, I packed my boat stern foremost this time so that I would sit in the bow seat, putting me closer to the center for better control.

Downriver, a rusty old sand dredge towered over the river on the high left bank. I clambered up to a vista of sand dunes backed by pines. The mine was apparently abandoned and being used as a four-wheeler track, judging by the fat, knobby tire prints. There are more than 250 sand and gravel surface mining operations in the Pascagoula basin, 50 of them here in Forrest County, according to DEQ.

Gravel mining improperly done can wreak havoc on streams. When river water pours into a low-lying gravel pit, the bed begins to collapse on the upstream side as gravel crumbles into it. As the ulcer gnaws its way upstream, the river spreads out and becomes shallow, a process known as headcutting. Pretty woodland streams can turn into broad, pebbly beds when riverside sand and gravel mining becomes widespread.

Some Pascagoula tributaries, including the Okatoma and the Bowie, have felt the effects of mining. In the 1990s, the DEQ fined a company $6,000 for mining near the Okatoma without a permit, and ordered the site cleaned up and replanted. The Bowie remains marred by gravel pits along its lower stretches. However, I haven't seen anything in the Pascagoula basin that can compare to the damage inflicted on the Bogue Chitto, Tangipahoa, and Amite rivers to the west. These streams run from southwest Mississippi into Louisiana, where there are extensive gravel operations. Aerial photos reveal the effects much like X-rays show the spread of cancer. On the Amite, a 1978 photo shows no gravel operations of any kind. By 1983 large mines have blossomed like tumors, and by 1994 once-small sandbars have turned into mini-Sahara Deserts. On the Tangipahoa, a forested stretch in 1967 became wide, sandy, and open by 1996. The headcutting on a tributary, Big Creek, moved upstream at a rate of 220 feet a year over 20 years. Damage is similar on the Bogue Chitto in Louisiana. However, when a Louisiana mining company set up an illegal operation on a Bogue Chitto gravel bar in Mississippi in 2002, residents reported it to the DEQ, which made the offenders restore the site and sent them packing.

Travis and I paddled side by side through seemingly endless cutover, young pine plantations, and thickets of willow and privet hedge. I could smell the pungent, resinous odor of wood chips from somewhere, a logging operation perhaps. Travis and I discussed the merits and negatives of clearcutting and agreed that for all its problems, it's at least better than suburbanization. When woods are cut they can grow back, but when houses and streets are built, they're usually there to stay.

A hawk, a wild turkey, and three does at the water's edge seemed to confirm our conclusion. I caught a whiff of otter, a distinctive, musky smell. A turtle underwater clung to the downstream side of a log, apparently waiting for something edible to wash by. A dead coyote lodged in tree limbs where high water had washed it. A pair of crows called, "Hee, haw, hee, haw."

After we passed the Sims Road bridge, we started seeing some real forest again—tall hickory, water oak, and sycamore trees. Fishing lines trailed from low branches. The river had widened enough to accommodate johnboats, like one we saw parked in a tributary. The weather stayed mild, sunny, and breezy, the wind alternately with us and against us. A large, white American egret flew upriver, and little killdee-type birds strutted stiff-legged on sandbars. I snacked on dried figs, yogurt-covered raisins, and pecans, sipping often from my water bottle. We weren't finding any clay-bank springs, so we stopped at a spring branch to get water. I used Polar Pure iodine crystals and Travis wielded a hand-pump filter topped off with chlorine. Neither tasted as good as fresh spring water, but both were better than dysentery. We couldn't know what might lie up at the headwaters of the branch: a dead cow? a manure-filled feed lot? a pristine meadow of wildflowers? I sipped from a clear brook years ago and, when my stomach started grumbling, followed it upstream to find a garbage dump at the side of a road.

Mosquitoes buzzed around us, a sign the weather was warming. An osprey circled, then dove, hit the water, and came up with something heavy in its talons. By way of contrast, a fist-sized kingfisher flew up, fluttered, and dove beak-first into the river. We couldn't tell if it caught anything. A rustic camphouse sheltered in the woods; it could pass for a frontier cabin were it not for the small black satellite dish out front. For all the scenery, this was a noisy stretch, with U.S. Highway 98 and a railroad paralleling the river not far south. At one point the railroad came so near the bank that we got to watch a train rumble past.

The river approached a huge curve, which the map identified as Scott's Bend. Camps clustered precariously on the left bank. One, which flew U.S. and Confederate flags from a rusty swingset, appeared poised to collapse from erosion. Another was halfway undermined. The river looped around a vast sandbar with more camps lined up behind it. These were a little more upscale, some with nautical themes like orange lifebuoys hanging from exterior walls. A volleyball net suggested this sandbar served as an unofficial community park.

Below the bend, I noticed a bunch of small trees with showy purple blossoms. They looked vaguely familiar. At first I thought they were wisteria vines choking some trees—I'd already seen some of the sweet-smelling lavendar blooms along the river—but these flowers were growing from the trees themselves. I found them in my tree identification book and felt a bit sheepish for not having recognized them as chinaberries. Like many southern kids, I'd played in chinaberry trees as a child. Kids liked to attach strips of inner tube between the branches and shoot the hard berries at each other, or make blowguns out of hollow elderberry stalks to propel the chinaberries through. But I didn't expect to see them on a river. Like so many nonnative species, chinaberries were planted as ornamentals and have elbowed their way into the wilds. Mimosa was a favorite backyard tree known for masses of pink or cream flowers. Farmers used privet as a livestock hedge. Chinese tallow makes for colorful fall foliage. Kudzu was mistakenly believed to stop erosion. All are now crowding out native vegetation.

We saw an osprey again, perhaps the same one we'd seen fishing earlier. Perched in a tree, it flew as we approached, giving us a close-up glimpse of its stunningly white chest and underwings. "That's the best view I've ever had of one," Travis said. I remembered canoeing under one in Georgia's Okefenokee Swamp that all but attacked me for getting near its nest. With 6-foot wingspreads and eagle-like features, ospreys are formidable creatures.

The first motorboaters we'd seen came churning up the river, a sunburned, bare-chested, beefy, muscular man standing at the wheel as two passengers hunched beside him. He slowed the bass boat so its wake wouldn't bounce us.

"Thanks," Travis said. "How y'all doing?"

"Fine" was the curt reply.

As we approached Mahned (MAH-nod) bridge, we saw our first swallow-tailed kite. This is a signature bird in southeast Mississippi, not normally found in other parts of the state. It has markings similar to an osprey, strikingly black and white, but with a sharply forked tail. And while ospreys soar and flap, the kite rarely moves its wings at all, skilled in riding air currents indefinitely on its 50-inch wingspread—hence the name kite, no doubt. It eats mainly lizards and snakes, and skims the water to drink and bathe.

Mahned bridge—said to be from the name Denham spelled backward—was an abandoned, rusty iron hulk, its eeriness exacerbated by a double murder that occurred here in 1995. A massive logjam was heaped against the pilings, which will someday collapse under the strain.

Not far downstream we stopped at the mouth of Tallahala Creek, which flowed even more swiftly than Bowie. Travis made coffee on a sandbar while I paddled up a short distance. I'd canoed the Tallahala before and found it quite pleasant—but that was downstream, not up. The 120-mile creek runs roughly side by side with a sister stream, Bogue Homa, at times barely 5 miles apart. Flowing through the town of Laurel, the Tallahala has a dark past of excessive pollution, but it's in far better shape than it used to be, according to the DEQ. Logging takes a toll, though, and the loggers don't always follow Best Management Practices. I've seen long stretches of cutover with no buffer zone to prevent siltation. In places where the forest is undisturbed, Tallahala abounds in beautiful mountain laurels.

Downriver from the Tallahala, the Leaf made a hairpin turn to the right. The river had apparently hit a vein of hard clay and had to find a softer route. Springs seeped from the sheer banks, while a high sandbar basked in deep shade on the other side. We filled a couple of bottles and made camp on the sandbar, which was perfectly positioned to catch a fine breeze blowing upriver. "I don't know if I've ever had one better," Travis said of the campsite. The sound of trickling springs filled the air as I mixed spiced rum with spring water in my enamel mug. Travis prepared lobster-flavored instant noodles with hickory-flavored tuna. "I think the lobster flavor is mostly monosodium glutamate and salt," he advised as we chowed down. But it was filling.

A motorboat glided slowly upriver, a thirtyish man in white golf shirt teaching a preteen boy to steer. We waved as they passed. After supper Travis paddled across to fill more jugs while I washed dishes. The father and son returned, propelled soundlessly by an electric trolling motor as they fished.

"Anything biting?" I asked.

The man looked up. "Sir?"

Sir! Was it the white in my beard? Or was he just unusually courteous? I didn't know whether to feel respectable or just old. I repeated my question.

"I had a strike a few minutes ago," he said.

"Fishing for bass?"

"Yeah. Where y'all from?"

"I'm from Liberty. He's from Magnolia." In typical rural parlance, I named the towns nearest where Travis and I reside. We both live way out in the country.

"Where'd y'all put in?" the man asked, chewing tobacco.

"Taylorsville. Above 84."

"That's a long way! I floated from Tallahala to New Augusta once," he said, referring to a distance of 4 miles. "Where you going?"

"The coast."

"Dang! Be careful."

"Thanks. We are." I nodded toward Travis, who stood in his boat filling jugs from the bank. "He's getting water from those springs. Where are y'all from?"

"Here. Petal," the fisherman said as they moved down the river.

Petal was 20 miles upriver, a day's travel by canoe. But by pickup truck it was no distance at all.

Travis finished his task, then fetched Eddie's video camera and paddled upstream to film the sunset while I read in *The Blue Bear* by Lynn Schooler, a nonfiction book by an Alaskan wilderness guide about a deadly quest to find the rare glacier bear. His descriptions of the northern wilds helped me appreciate these southern scenes, like the way the sunlight receded from the bright clay banks as the water turned progressively deeper shades of green and gold.

Travis returned in the dusk and I made hot chocolate. We built a fire and he suggested I play some guitar, but I had torn my right index fingernail and it hurt to play. Ever resourceful, Travis suggested cutting a pick out of my plastic bailing scoop. It worked, albeit clumsily. I plucked a simple version of Bach's "Jesu, Joy of Man's Desire."

"What's the name of that piece?" Travis asked when I finished. When I told him, he said, "I've had that tune in my head all day."

Creeks like this add to the Pascagoula basin's dimension.

6

The river was bliss when we started out the next morning, but we knew it wouldn't last long because we were approaching the Georgia-Pacific paper mill at New Augusta. For now, though, I focused on enjoying the wide, slow river. Reflections of woods and sky forged the surface into a green-blue pane of glass. When I looked over my left shoulder, I saw a series of molten circles in the water from my paddle strokes. The outside corner of the paddle created a widening clockwise spiral, and the correction stroke sent a cross wave that converted the circle into an S. A dragonfly rode my shoulder, perching on the faded green fabric of my shirt. The air, sweetened with the honeysuckle smells of morning forest, felt cool on my right cheek as I paddled on the left side.

The sweet smells were soon supplanted by the raw odor of the paper mill, however, which grew by stages from pungent to putrid to vomitous. Birdsong faded beneath the clanking, clanging, and grinding of trucks, trains, and other motors. A sandbar held an old red gasoline can and a deposit of toilet paper and feces.

Past the New Augusta bridge to our left, an inflow pipe led from the sluggish river under a sign: "Caution: Occasionally turbulent water." It certainly wasn't turbulent now. A mobile home across the river with a dock didn't strike me as the best place for a camp. But woods and other vegetation hid the mill, and as we paddled, the smell began to dissipate more quickly than I expected. Altogether our encounter with the mill was anticlimactic from a place that prompted a flurry of lawsuits in the 1990s claiming dioxin contamination, all ultimately dismissed.

The paper mill merely reflects land use throughout the Pascagoula basin, 59 percent of which is in trees. Of that, 22 percent is owned by corporations, 14 percent by the government such as national forest, and the rest is in private landowners' hands. At the turn of the 21st century, the Pascagoula basin was producing more than a quarter of the state's timber income.

The G-P plant is one of 143 industrial dischargers permitted in the basin by the DEQ.

Timber has been a commodity in the Pascagoula basin for centuries. Early settlers not only cut trees for lumber but used them to make charcoal, pitch, and turpentine. To produce charcoal, workers stacked logs and covered them in dirt, then baked them to reduce the wood to nearly pure carbon used for fuel. For pitch, workers notched the trees and gathered the "gum" for such naval uses as caulking ships' hulls and tarring rope. Turpentine resulted from cooking the gum in copper stills. The acrid spirit went into paint, varnish, ink, and plastics. Sometimes at Christmas people poured barrels of turpentine on the water's surface and set it afire for primitive fireworks. They also warded off redbugs and ticks by tying turpentine-soaked rags around their ankles.

Travis and I snacked on oranges and chatted, then I pulled ahead. An osprey whistled and swooped off a limb, carrying a fish. I had disturbed its breakfast. It perched again, then flew once more as I neared, the fish hanging from its talons like a payload on a bomber. The next time I saw the big raptor, it had presumably finished eating and swooped to a treetop to watch this rude intruder pass underneath.

A large sandbar to our right served as a front yard for a subdivision of homes and camps half-concealed among the trees. Across the river, Bogue Homa Creek entered. I had canoed the creek before and found it beautiful but choked with logs due to its small size. We turned up it and stopped at a sandbar. Travis didn't want to fight the current but I was eager to probe upstream, so I left him and paddled around a corner to a fallen oak tree that halted passage. I beached the boat and found a perfect sandbar just beyond the fallen tree, with overhanging limbs providing shade and a waist-deep pool of clear, amber-colored water in the sunshine. We were ahead of schedule, so I thought perhaps we should make an early camp and take a day to rest and clean up. On the other hand, it was only 10 a.m., and I wasn't sure I wanted to squander so much paddling time.

I had decided to keep going when a great blue heron swooped down and landed just across the creek. When it saw me it flew off, but the fact that it stopped at all was remarkable. It's rare that the sharp-eyed herons come so close to humans. Five months earlier I had ignored a similar appearance by my spirit animal on Texas's Village Creek and wound up regretting it.

On that float, a heron lighted on a secluded sandbar right in front of us and walked around as if recommending it. Since it was just 2:30 p.m., I ignored the invitation and went on, only to enter a long stretch of camps, houses, roads, trucks, and docked motorboats. We were forced to camp in a state park where mobs of campers, including a troop of Boy Scouts, kept us awake most of the night. To make matters worse, the next morning I discovered someone had stolen the battery out of my truck, and when I bought another one I was overcharged by $10, I later learned. If only I had paid attention to my spirit animal, we would have had a peaceful night's sleep, I would have remained in harmony with the universe, and I'll bet I wouldn't have been burglarized, either. So when the heron gave me a sign this time, I took heed.

While waiting for Travis to catch up, I stripped and waded into the breathtakingly cold water, washing from head to toe with biodegradable soap. I hadn't bathed in a week. I scrubbed my sweat-caked clothes, putting on my one set of clean ones. Travis still hadn't showed and I was just getting ready to go fetch him when he appeared. When I suggested camping here, he was skeptical. But it was noon so we ate lunch and he slung his hammock for a nap, and that put him in a mood to stay. Meanwhile I read, jotted notes, and dozed. In midafternoon a swallow-tailed kite appeared directly overhead, not too high. I looked straight up at it and saw it clean its talons with its beak in midflight. Its chest was as white as a dress shirt against a gray tuxedo.

What a place! It was apparently inaccessible: too shallow to come up in a motorboat, too logjammed to float down in a canoe, and no roads or trails leading to our camp. A pleasant breeze blew up the creek, carrying the scents of deep woods. Our canoes rested snugly in sandy shallows, and my clothes dried on logs and branches in the sunshine. This was a Garden of Eden. Well, almost: now and then we heard a roar to the south, the sound of some sort of artillery at Camp Shelby, a National Guard base.

After a few hours of lounging, I needed exercise, so I put on shorts and wading shoes and set out up the creek. Progress involved picking my way over submerged logs, bellying into cold water, then emerging onto sun-scalded sandbars. Each sandy expanse told a tale of animal passage. Beaver trails showed big webbed feet on either side of broad tail drags. Raccoon tracks resembled little human hands. Great blue heron footprints stretched nearly as long as my own. Mink, muskrat, snake, armadillo, deer, wild hog, canine, squirrel—nearly every indigenous species seemed to have left its sign on the white sands of Bogue Homa.

Fish darted in the clear water, and for the first time I wished I'd brought some tackle. I have found fishing to be impractical on long downriver trips, but it sure would be nice right now to toss a lure into a deep pocket to try to catch a mess of fish. They'd be mighty tasty grilled over the coals, as opposed to the rice and tuna I planned to cook for supper.

Back at camp I sat and examined gravel pebbles. When wet they became lustrous, polished black to porous white and everything in-between: grays, oranges, browns, lavender, pink, coral, brick, yellow, beige. They came in all shapes, most smaller than a quarter.

Across the creek I noticed a fallen log that resembled some sort of big predator, a wolf or panther, striding out of the woods to the water, paw forward. It was so vivid it seemed to have a spiritual presence. Perhaps it would come alive in the night. It was easy to see why native Americans were animists, believing spirits inhabited natural objects.

An osprey flew up the creek, glancing down at us from 15 feet high. At sunset three kites put on an aerobatic display, soaring in circles and figure eights over the treetops. Three egrets winged toward their roost. Flocks of small birds with brown wings and yellow bellies flashed by whistling, too fast to identify. A squirrel chirruped mournfully. Bats darted overhead as dusk settled in. Owls hooted; frogs screeched. I built a fire, and a beaver slapped the water in the dark a few feet away, making me jump.

This deep-woods creek was another side to the Pascagoula basin, one not seen from the big rivers. Creeks like this make up the circulatory system of the Pascagoula River system. There are 15,045 miles of streams in the basin's 9,600 square miles, according to the DEQ. My longtime boss at the newspaper, Charles Dunagin, 69, grew up on some of the smaller waterways.

"My brother and I used to swim in Providence Creek," he told me later, referring to a tributary to Bowie Creek. "In the summertime it was an everyday affair. That was where we took a bath. They didn't have any indoor plumbing out there." Reece's Creek, Lott's Creek, Bowie, Bogue Homa—Dunagin knew them well. He avoided Tallahala Creek, though, because of its reputation.

"He [Dad] always discouraged us from getting on the Tallahala because he said it was polluted," said Dunagin, pronouncing it "Tal-a-HAY-lee" instead of "Tal-a-HAY-la." A Masonite plant and the city of Laurel both used the creek as a dumping ground. "These streams were not as clean then as they are now," said Dunagin, who became a member of the Mississippi Commission on Environmental Quality after his semiretirement at the McComb *Enterprise-Journal.* The Hercules chemical plant deposited waste in a tributary to Bowie Creek, Dunagin remembered. "I had an uncle that had a dog, and if that dog got fleas on him, they'd take and put him in that creek that Hercules dumped in. I don't know how long that dog lived, but it killed the fleas. They dipped him in that creek."

Dunagin learned to fish on Bogue Homa not far from where we were camped. "My dad always encouraged us to go to the Bogue Homa because it was cleaner," he said. One of his most memorable trips as a boy was to the mouth of the creek, where he and his uncle got into a bream bed and the fish bit like crazy. "I really got hooked on fishing early on," Dunagin said. "Then we started floating the Leaf some. Usually we'd put in at the mouth of Tallahala and float to the Old Augusta bridge."

One day the Leaf River washed up a 10-foot wooden skiff, and Dunagin and his friends decided to have an adventure. "We were going to meet up there and camp and set out hooks on the Leaf River," he reminisced with a smile. "This guy that owned the boat was taking supplies across the river to a sandbar. He had taken his little brother over there and left him. The boat got caught in the current and turned over. The boat washed away then. The little brother was stranded on the other side of the river." Dunagin's friend swam across and led his brother through the woods to a road, where Dunagin picked them up in his car. "I had a 1936 Ford with the back end cut out that I had swapped a calf for. It was a real clunker, but you could go a few miles without the radiator conking out." As for the wooden boat, "He had painted the boat and called it *Old Faithful.* His daddy said he should have called it the *Titanic.*"

Around 1960 Dunagin's father, grandfather, and brother undertook an ambitious project. "They built a houseboat at Old Augusta, built it on 50-gallon drums that Dad had got at Hercules where he worked. They floated it down to Cumbest Bluff and left it there for awhile, then took it down to Poticaw Bayou," Dunagin said, referring to a journey of more than 100 miles to sites on the lower Pascagoula River. "They may have had a motorboat that went down beside them. Mostly they got some big poles and poled down. I went down there and picked them up." His dad kept a houseboat at Poticaw Landing, and they often fished together there.

I got a good night's sleep in our Bogue Homa camp. For the first night I wasn't disturbed by traffic or train noises, just the roar of night creatures in the enveloping forest, which to me is soothing. We broke camp early, eager to make up for yesterday's lost time, though in fact we were right on schedule in terms of mileage. But as we emerged from Bogue Homa onto the Leaf, we encountered a white wall of fog. I don't recall ever seeing one so thick on a river. It was spooky and disorienting. I couldn't tell where the water ended and the air began. At one point I reached too far with the paddle and almost lost my balance, making me momentarily dizzy. Now and then I could see black shapes of logs or branches protruding from the water. "I thought I saw Nessie a couple times back there," Travis commented, meaning Scotland's legendary Loch Ness Monster.

A pair of tents sagged forlornly on a sandbar; either they'd been pitched for days and not tightened up, or whoever set them up didn't know what they were doing. Farther on we passed a man loading something onto an idling four-wheeler at the water's edge, and a couple with a truck backed down to the river's edge preparing to launch a boat—all of this in a thick white mist. I said good morning and the woman replied tersely. She had long dark hair, a flannel shirt, and ragged jeans. They asked Travis where we were going, and when he said the coast the man just laughed.

I stopped on a sandbank to adjust my load. With no headwinds, I had switched back to the more comfortable stern seat. The couple motored by in the fog. I waited for Travis, hoping he wouldn't pass me unknowingly. I blew my nose and he emerged from the fog across the river. He was indeed in the act of passing me, neither of us having seen the other in the mist. That would have been a problem. I got separated like that one time from Scott at night on the Pascagoula River and couldn't determine if he was ahead or behind. I didn't know whether to wait for him to catch up or paddle fast to catch him. I decided to wait, which was the right guess, for he eventually paddled up from behind, but getting separated in such circumstances is dicey.

As the fog dissipated, the river stretched out smooth as a lake, trees looming on each side. An immature little blue heron flew by. Otherwise there was little sign of life. The air was damp and chilly enough for me to keep on my wool shirt, yet the sun behind the fog was glary enough to make me wear my sunglasses.

We saw more motorboaters, no surprise since this was a Saturday. A group of three men was checking a trotline. "Morning!" one of them called. Then he said to a boatmate, "Give me a beer." It was 9 a.m. A little later they passed us heading downstream.

The river presented us with a few swift stretches, but mostly it was like paddling a north-woods lake. One high, forested bank a short distance above Beaumont provided a scenic spot for several camps nestled among the trees. Wooden steps protruded down the bank from one camphouse with a sign advertising a real estate company. The current had undermined the steps, and a new path had been forged down the steep sandbank. The sand had washed into the river there to form a shoal on which I almost grounded. For me the scene was more evidence that even the most innocuous use of a river has a domino effect. This was a tasteful camp in the woods on a bluff, yet a simple wooden staircase precipitated a shoal that affected the river channel.

At Beaumont, a steel staircase led up a high bank to D&B Bait Shop. Travis and I walked up to find the three beer-drinking fishermen loading their truck. They told us they had 32 fish, and showed off a couple of big spotted cat. The men had fresh drinks and red eyes. No doubt they'd been running lines all night.

The store was a humble plywood shack manned by a middle-aged couple who eyed us glumly as we perused a small selection of chips and Nabs. An old, white, upright refrigerator held cans of RC Cola, A&W Creme Soda, Nehi Orange Crush, and Yoo Hoo chocolate milk drink. Bait and beer were obviously the main sellers here. I took a creme soda and Nabs, Travis a bag of barbecue chips.

"Where'd you put in?" the proprietor asked after we paid. When Travis said Taylorsville, he commented, "Going to make a night of it, huh? How far you going?" When Travis said Gautier, the man responded, "Y'all are going to make a weekend of it." Back on the river, Travis and I couldn't agree on whether the man was being ironic or just ignorant of regional geography. I maintained the latter, but Travis was convinced he was making deliberate use of understatement.

Just below Beaumont we passed Thompson Creek, which looks impressive on a topo map—long and squiggly—but is barely big enough to wade even at the mouth. Soon after passing the creek, to our surprise we found more clay banks and springs; we thought we'd left all that behind on the upper sections. But a gatory splash from a willow thicket made it plain we were approaching swamp country.

Two swallow-tailed kites flew in synchronization as we ate lunch on a damp clay shelf. Travis said he'd seen about 10 of the birds today. I scoffed but believed him. He was more attentive to wildlife, especially birds, than I—a result of his slower speed. I frequently had to stop and wait for him to come into sight behind me. At lunch, by which time we'd already made 15 miles, he said, "I'd like to throttle back the pace a little bit this afternoon. Seems like it's all zipping past me." I told him we weren't going nearly as fast as I'd prefer: if we didn't have to rendezvous with Scott, we'd be halfway down the main Pascagoula by now. As usual, having to accommodate Scott was holding me back, all so he could make money counting snags.

Motorboat number five passed, a young couple. I decided to let Travis get a head start after lunch. After he left, I took notes and admired the scenery. A few puffy white clouds drifted high overhead, the first we'd seen in days. The breeze was still cool but increasingly coming from the south. Finally I set out, trying to make myself go slower. It's just not in my nature. I did stop paddling long enough to speak to a man in a boat with three kids. When I asked his luck he held up a stringer of small bluecat.

In midafternoon I found Travis sitting in the shade sipping coffee and reading Steinbeck. I pulled up, stretched out on my life vest, and took a nap—I wasn't too rushed for that—then we continued on together, passing a motorboat weighted down with three obese teenaged boys. An old iron bridge at McLain was closed to traffic, but a new bridge was under construction just beyond. We checked out a big sandbar across from a high ridge that the map called Dee Dee's Hills, elevation 155 feet, but found it mottled with four-wheeler tracks, so we kept plugging along. At a huge sandbar before the Highway 98 bridge, a teen on a four-wheeler was trying to pull another boy on skis. Down the beach, families played and swam.

The wind picked up, right in our faces. Progress became a fight. Gusts blew my hat brim back and whipped the words out of my mouth as I called to Travis: "I'll bet we see a lot of this on the Pascagoula."

Bridge at Merrill looms out of morning fog.

7

The river angled east, giving us some protection from the headwind. But there were other obstacles, like a sand shoal that appeared in front of my boat, barely an inch deep and halfway across the river. I made a hard left turn and stroked across to the far side. As I neared the log-strewn shore, one of the "logs" stretched out and slid into the water: an alligator, 8 feet long, its black hide coated with pale dirt. The gator's underwater trajectory put it close to the bottom of my canoe. I looked down to see if I could spot it in the murky water but couldn't. I'd had enough encounters with gators not to worry, but there's always something unnerving about a reptile larger than myself.

Kingfishers chittered overhead as we headed for a shady spot at the end of a long sandbar in hopes of making camp. But we found tire tracks and footprints, so I set out across the river to investigate another bar. I aimed my canoe between a pair of stobs in midstream, discovering too late that they both belonged to the same barely submerged log. Skimming over it put me broadside against a stump in swift water, a situation ripe for capsize. I slid the boat off just in time, heart thumping. It's that easy to come to grief in a river.

The sandbar was high and long, backed by thick grass and woods with no apparent access. But as I strolled along the upper edge, I spotted a tent and a little boy hammering something farther down the bar. Those folks were probably boat-camping. Without a word I returned to my canoe and paddled back across to Travis. It was too late to keep going so we decided to stay, despite the tire tracks. They looked old, anyway.

We pitched our tents on a long, narrow, shady beach backed by pungent willow forest. That smell, though not especially pleasant in its own right, is full of sweet memories for me. It conjures up boyhood visits to the Mississippi River and manhood canoe trips down countless southern streams. Travis went exploring on foot and returned to tell me about some yellow wildflowers he had found. I followed him up into the woods to discover a luminous, waist-deep understory of gold. The flower petals were outnumbered only by mosquitoes. The sight was glorious, the misery exquisite. As usual, we couldn't find the flowers in our field guide, but someone told us later that locally they're called yellowtop.

Back at camp, Travis treated us to instant noodles with packaged salmon for supper. With enough imagination I could pretend we were at a Chinese restaurant in Anchorage. There was no firewood around, but we didn't need a fire on this pleasantly cool night. We reclined on the beach watching the distant blinking lights of a radio antenna, the breeze keeping mosquitoes away. The other campers were out of sight on the far side of their sandbar, but we could see the radiant glow of their campfire—bonfire, more like it—plus the electric white circle of a Q-beam playing across the far bank.

The topic turned to canoe-camping. Travis chided me gently for my traveling habits, saying he doesn't like to "linger" at camp in the mornings like I do. What he didn't realize was the only reason I lingered was to keep up with my notes for this book. I'm usually the one who gives my partners a hard time for sleeping late or dallying over morning coffee. But I didn't argue. I knew the hand of karma when I saw it. It was my turn to suffer the scorn of the young and energetic.

We woke to another blinding fog. Everything was wet, and dew sprinkled us like tree sap. A turkey gobble sounded wooden, probably a hunter. Travis redeemed himself for his weird suppers by cooking pancakes served with honey, a real improvement over the usual instant oatmeal or cold cereal. Walking the sandbar after breakfast, I noticed it was crisscrossed with mouse tracks. The little fellow had gone from shelter to shelter—a hole under a log, a clump of willows—in its midnight rambles. At risk from snakes, owls, foxes, and bobcats, it faced as many dangers as Ulysses in *The Odyssey*.

Travis and I paddled side by side in the fog in order not to lose each other. We passed the campers: a few adults and a bunch of children with two tents and two johnboats. A mile or so farther on we saw another encampment—tent, people, a boat—making a gray-black silhouette in the fog.

"Where y'all going?" a man called from the breakfast fire.

"The coast," I replied.

"Whooeee! Y'all got a hike ahead of you."

"A good hike," I countered.

A deserted sandbar harbored a wild turkey, which ran and then flew into the jungle-like woods when our boats surprised it. Spider webs draped from logs like fish nets, outlined in mist droplets. The fog slowly lifted, bringing color to the world. Shoreside trees, green and gold with sunrise, stood out in sharp contrast against the misty jungle. "It looks like we're going to see Colonel Kurtz or something," Travis said.

In Joseph Conrad's *Heart of Darkness*, the narrator worried constantly about damaging his boat. I had the same problem with my creaky old vessel. My fears were not unfounded: I was rounding a bend when a seat screw snapped. We pulled over to shore to examine the damage. I'd had reservations about taking this old warhorse on such a long campaign. Downriver trips with gear put a lot of strain on a boat, testing every joint. But good old Travis produced a repair kit and I wired the seat back in place, functional if none too sturdy.

We passed several camps, some rustically attractive, others with massive homemade erosion control efforts, like a steel grid paved with tin. Such devices rarely last long, and this one was already half-destroyed by logs, a real eyesore.

A small diamondback water snake, nonpoisonous, was swimming the river when it saw my boat. It stopped and wiggled its tongue. Travis mentioned seeing several turtles he couldn't identify, reminding me that there are species found only in the Pascagoula and nearby Pearl river basins, such as the yellow-blotched map turtle.

We came to our first real backwater in a bend to the left. Eager to see some swamp, we paddled up into a series of shallow, muddy coves ringed with willows and resounding with bullfrog croaks like warm-up time at a tuba convention. As my paddle probed the bottom, the mud exhaled a sulfurous stink. We picked our way through the coves and returned to the river, stopping for lunch on a sandbank in the thin mesh of willow shade. We were just 2 or 3 miles from the juncture of the Leaf and Chickasawhay, and only 1 bend upriver from a spot where, according to our maps, the rivers nearly touch. Travis was eager to walk overland there and see the Chickasawhay, which shouldn't be more than 100 yards away.

When we reached the spot, we found a bayou leading east, presumably connecting the two rivers. But a fisherman was angling there and we didn't want to disturb him, so we landed just upriver and climbed a steep bank into the woods. A logging road led east and then north, revealing nothing but a cypress swamp where the Chickasawhay should be. We returned to the river and floated just below the connecting bayou, again climbing a steep bank and hiking through the woods. Now we found a big bayou instead of the Chickasawhay. The topography was at odds with our maps. I later learned why: the Chickasawhay had cut through a bend upstream of here, bypassing the stretch that had once nearly touched the Leaf. What we thought was an old bayou was in fact the old Chickasawhay channel, now a dead lake.

As we paddled toward the juncture of the rivers, I marveled at the white sand, green forest, blue sky, and reflective water. Sooner than I expected, I saw a big iron bridge up ahead. It had to be Merrill, yet we hadn't even reached the Chickasawhay, and Merrill lies half a mile or more below the juncture. As we neared, though, I realized that what looked like a downstream reach of the Leaf's left bank was in fact the far side of the Chickasawhay as the two rivers ran together, a riverine optical illusion.

A sand point protruded at the union. As I swung around and paddled up toward it, I noticed a distinct difference between the Leaf and the Chickasawhay. The former flowed dark and clear, the latter pale and silty. With my canoe positioned lengthwise along the line of demarcation, I stopped in the shallows and stuck a foot in each river: my left in the Leaf, right in the Chickasawhay, the boat hull on the Pascagoula. I felt the pulse of the South, a fine double power sweeping down from the forests and hills of Mississippi.

The word Pascagoula is the name of an Indian tribe and means "bread people," referring to a staple of their diet. The name Leaf River is an English translation of the Choctaw word *hastaphatcha*, which means "river of falling leaves," according to anthropologist Chris McPhail, who analyzed place-name origins in his 2002 University of Southern Mississippi paper "Mississippi Rivers: A Study of Choctaw Indian Place-Names of the Streams and Rivers of the State of Mississippi." Chickasawhay was a subgroup of Choctaw Indians and the name of a village where they lived; the word means "place where martins dance," which McPhail believes refers to a high river bank, King's Bluff, where the birds gathered.

Here at the juncture, French explorers found an Indian town called Les Capinans. Now, some 3,280 acres between the rivers and east of the

Chickasawhay make up the Nature Conservancy's Charles M. Deaton Preserve, named for a former state legislator and chairman of the Mississippi Wildlife Heritage Commission instrumental in preserving the Pascagoula. Deaton, of Greenwood, also helped establish Dahomey National Wildlife Refuge in Bolivar County, Clark Creek Natural Area in Wilkinson County, Hillside National Wildlife Refuge in Holmes County, and Sweetbay Bogs in Stone County—quite a legacy. Just downstream on the west bank of the Pascagoula lay the Conservancy's 1,692-acre Herman Murrah Preserve, named for a late game warden and Pascagoula Hardwood Company caretaker also involved in the Pascagoula preservation efforts. "He had lived in the swamp all his life and probably knew it better than any other man alive," says Donald Schueler's *Preserving the Pascagoula*. The two preserves and the Pascagoula River and Ward Bayou Wildlife Management Areas join to create an 80-mile natural corridor along the river.

We turned our boats south toward Merrill. We were in the Pascagoula River proper now, wider and more windswept than its two main branches. The Leaf and Chickasawhay eddied together, forming a long sand shoal along the east side which in turn forged back-currents that slowed our progress. We bounced under the old iron bridge, built in 1928 to take the place of a ferry that had operated since 1889. Just past the bridge stretched a rail trestle. The town of Merrill was named for Colonel Frank B. Merrill of Rochester, New York, who spearheaded the effort to build a rail line connecting Mobile, Alabama, with Kansas City, Missouri, in 1898.

Merrill has two boat ramps on the east bank: a private landing between the two bridges, and a public landing just below the railroad. We had permission to take out at the former. I swung my boat around just past the ramp, careful not to scrape its fiberglass hull on the concrete, and tied to a wooden post. Then I climbed out, stiff-legged and sweaty in my long pants, long-sleeved shirt, rubber boots, and wide-brimmed hat. I took a long pull of warm water from a Clorox jug. Travis and I walked up to a grassy area in front of a massive wooden camphouse on stilts shaded by big live oaks humming with cicadas. Camp owner Ransom Jones, who wasn't present, had given us permission to store our boats and gear under the building if necessary. We walked a couple hundred yards up a gravel road to Nature Conservancy headquarters. The renovated camphouse was neat, green, on stilts, surrounded by blooming Indian hawthorn shrubs and shaded by a massive live oak. The house fronted Main Street, a fancy name for a narrow road with a few camps on it.

Becky Stowe, who manages the two Nature Conservancy preserves, wasn't here since this was Sunday, but she had left a key in the mailbox for us, and we went inside. She had invited us to stay in the house and use the shower. Travis and I certainly planned to do the latter, but we decided to camp near the river since it was too far to haul our gear; plus we didn't feel like sleeping indoors. Camp life will do that to you.

After a nap and a stroll, I met some fishermen taking their boat out at the ramp. When I mentioned our trip, one quipped, "You must like canoeing a whole lot." They told us, to our surprise, that they had seen Scott camped just four bends up the Chickasawhay. We had planned to rendezvous with Scott here tomorrow, and I had no way to let him know we had arrived a day early. We didn't leave our cell phones on, to keep from running the batteries down, but we had agreed for him to call me at 6 this evening.

Eddie McCalip, who planned to film this midpoint in our journey, arrived with his wife, Tina, and 9-year-old son, Tyler. I paddled Eddie upriver to the forks as he filmed, then fought the wind back down to the boat ramp. Eddie asked me to take Tyler for a spin, so off we went back upriver. When I turned around, the wind had become so strong I could barely make it to the landing, in large part because I had taken out most of my equipment, making the boat easier for the wind to shove around. The experience convinced me to switch ends of the canoe again tomorrow; my jerry-rigged seat was wobbly anyway.

Scott called at 6, surprised to learn we were at Merrill. Since he had already made camp, he told us he'd see us tomorrow rather than come on now. TV man Walt Grayson was also visiting at noon to film us, he said. With all this media attention I was beginning to feel like a celebrity, albeit a hot, dirty one.

I treated Travis to a gourmet dinner of rice with minestrone soup mix and hickory-smoked tuna. The sun set behind the iron bridge as flocks of white egrets flapped west. The wind died and the river curled like liquid glass. Dusk settled in as Travis and I swatted mosquitoes and talked. Recalling the fisherman who'd told us we must like canoeing a whole lot, I said, "That made me wonder if I really do like it a whole lot."

Travis replied, “One thing I’ve learned from this trip is I really do like canoeing a whole lot.”

The exchange made me feel jaded. Perhaps it was only in contrast with my 29-year-old companion, who looked forward to the many journeys ahead of him—some of which I’d already done. It occurred to me that I’d pretty much been everywhere I wanted to go, canoed all the rivers I wanted to canoe. Did that mean I’m ready for a rocking chair and a grandbaby on my knee? Well, yes—I have two grandkids and a great front porch—but not only that, of course. Heck, I felt as strong as ever, more so since I have the weight of experience and no longer waste energy testing myself. I’ve learned that my pleasure in canoeing comes mostly from exploring. I’d rather paddle a mediocre river that’s new to me than a great one I’ve done repeatedly. True, I had floated long sections of the Leaf, Chickasawhay, and Pascagoula previously, but never in one wallop.

I showered, then walked up Main Street toward the bridge. Even though it was April, yellow Christmas lights festooned a house on stilts across the road, outlining the roof and handrails. The scent of evergreen trees made me think of Montana, specifically Glacier National Park, where I had gone backcountry camping the previous summer. In some ways this place felt even more remote than there. As I walked out onto the steel mesh floor of the bridge, I couldn’t help but feel vulnerable and a bit spooked.

Returning to the boat ramp to admire the stars, I made out Orion, what I thought to be Cetus the whale, and a bright planet in the west. I have even less knowledge of astronomy than of natural history. A car crossed the bridge from east to west, its headlights bringing the iron structure of diamond and triangular shapes to silvery life, then erasing them with instant darkness as it passed. Red taillights disappeared into the woods across the river.

The area around the ramp was pleasantly dark. Up where we had pitched our tents was not. We had unwittingly placed them under a sodium security light. I zipped the tent door shut and lay facing away from the glare, but I tossed and turned anyway, muscles restless and achy. I dreamed I visited some neighbors, who acted aloof toward me, which hurt my feelings. It occurred to me later that the dream must have come from the overall lack of conviviality we had found along the river so far. Most people had been terse, if not unfriendly. After giving it some thought, I theorized that it was because we were paddling in solo boats, making us seem a bit more menacing, or at least inscrutable. On trips where I shared a canoe with a companion, people acted more hospitable.

Morning mist was so thick I couldn't see across the river. The bridge arched away into the fog as if to another dimension. Dew drenched everything. I breakfasted on cold cereal and hot tea, then sat and read at the edge of the concrete slab under the camphouse, the only place relatively dew-free. I heard a noise at the boat ramp. Odd. I resumed reading. Then someone called, "Hey, where do you get some good coffee around here?"

Scott.

The Chickasawhay is narrow and swift at its headwaters near Enterprise.

PART TWO

CHICKASAWHAY RIVER

SCOTT B. WILLIAMS

Drinking water can often be found in springs that run from bluffs like this overhanging the river.

8

A flash of white light followed by a resounding crash of tree-shaking thunder jolted me out of peaceful sleep on my Therm-a-Rest mattress. I sat up in the blackness of the tent as cognizance of my whereabouts returned with wakefulness. Heavy dollops of rain splattered off branches and onto the tent fly in the wake of the thunder. I groped about in the dark for the mesh pocket sewn into the sidewall of the tent, and found my Mini Maglite. Shining it onto my watch face, I confirmed my suspicion that it was still early—just 3 a.m.—way too early to break camp and get moving.

I felt trapped as I sat there weather-bound, high atop a steep bank overlooking the Chickasawhay River. The woods here were so thick I had to hack out a space to put the tent the evening before with my machete. Travel in such a forest by any means other than the river would be difficult even in full daylight. To attempt it at night would require desperate circumstances—something like a snake bite or other medical emergency, at least.

It's not that I felt any need to go anywhere. I had everything I needed for the next two weeks either inside the tent with me, or just outside, securely stowed away in the watertight compartments of my 17-foot kayak. But the knowledge in these situations that I couldn't simply leave at will was at one time quite unsettling, at least in the earlier years of my solo kayaking, canoeing, and backpacking excursions. It takes some time to get used to the idea of moving according to nature's rhythms when one grows up in a society where arbitrary time designations dictate every action and activity. And even for me, one who has put away my watch for weeks and even months at a time on previous trips, the first night back out in the *real* world is always an uncomfortable adjustment.

Rain on a river trip goes along with the territory, and normally would warrant little concern, other than the added hassle of making and breaking camp while trying to keep gear and supplies dry. But on this trip, rain on the first night and a forecast of plenty more to come had me quite worried, because this time I had work to do that could not be performed in anything beyond a light drizzle. I wasn't even supposed to be on this river, except for a last-minute change of plans following the commitment to do this work. And now, sitting in pitch-black darkness as the rain pounded my lightweight roof, I worried about rising water levels, impossible conditions for photography, and the possibility of losing the extra money I planned to make in the coming days.

I was less than half a mile from the head of the Chickasawhay River, at the place where the Chunky River and Okatibbee Creek run together. Downstream of my bush campsite, the Chickasawhay coursed 159 miles to its meeting place with the Leaf. I planned to be at that other confluence in 9 days, when Ernest Herndon and Travis Easley were supposed to finish their shorter run down the Leaf. But if this rain continued, I would have to sit it out. I had data to gather, and photographic evidence to collect. Notes had to be taken, and over a hundred pages of documentation I carried in a waterproof bag had to be verified as I worked my way mile by mile downriver. None of this could be done in the rain.

I thought of Ernest and Travis over on that other river, a sister to the Chickasawhay, and thought how easy they had it. With nothing to do but paddle, they could ease down the river, enjoying the scenery, cooking good meals, and sharing campfire conversation. The rain wouldn't be much of a bother to them. They'd pull on ponchos or rain jackets and mosey on downstream regardless, making an easy pace of about 13 miles a day to reach the arranged meeting place. Plus, they had a 3-hour head start when they put in above Taylorsville the day before at around 1:30 p.m. and probably made several miles on Saturday. I, on the other hand, because we were all riding in the same truck, had to wait on them to launch and then ride another two hours to my launch site at Enterprise. It was 4:30 before I even had my boat in the water, and then I had to paddle *upstream* to get to my starting point, the official beginning of the Chickasawhay River. They weren't starting anywhere near the beginning of the Leaf, just the beginning of easy paddling where the river is big enough that there are no logjams or significant shoals to force a river traveler to have to get out of the boat. Once again, I'd let Ernest manipulate me into helping him fulfill a dream that was his idea, not mine.

Ernest has had a way of doing that ever since I met him back in 1986. I don't know if it is because he is several years older, and more experienced in the ways of coercion and persuasion, or if it is simply my fault for being too easygoing and agreeable. It is likely a combination of both. I met Ernest when his wife, Angelyn, signed up for a self-defense class that I was teaching near their hometown back when martial arts, rather than boating and exploring, was my obsession. In my early twenties at the time, I had not had many opportunities to learn to be an adventurer growing up in a small town like Prentiss. But I had been reading *National Geographic* since childhood, as well as countless novels and travelogues set in jungles, mountains, and on the sea. So when Ernest told me he had been on two long bush patrols into the heart of equatorial New Guinea, I was understandably impressed. And he was a writer—a newspaper journalist with his first book, an account of one of his New Guinea adventures, about to be released. So this is how he managed to get the upper hand on all of our subsequent adventures from the very beginning. He was just looking for some younger, impressionable, would-be adventurer like me, so he would have a travel companion that he could mold to his ideas and persuade to take the trips he envisioned. Unlike me, Ernest does not like to travel alone. And unfortunately for him, now that I've wised up to his ways, there's no way I would spend two weeks on the same river with him. Hence the reason he was now starting down the Leaf with 29-year-old Travis Easley—another young, impressionable, would-be adventurer he can mold to his ideas . . .

But even though I was not on the same river with him, the fact that I was once again on a river with a boat full of camping gear was a testament to his persuasive powers. I started my first tentative forays into the backcountry in a used fiberglass canoe. I progressed through an assortment of canoes and sea kayaks over the years and paddled literally thousands of miles, one time 2,600 miles on a single trip, from Canada to Vicksburg. Canoeing and kayaking were like everything else I've tried in life; to me one of the greatest pleasures is learning and perfecting new skills, then moving on to whatever next level my interest leads me. With all these river miles under my keel in paddle-powered boats, my next endeavor was to learn to build small boats with modern wood and fiberglass construction techniques. This naturally led to a broader spectrum of boating interests, and my attention turned to sailing rather than paddling. I built a couple of small sailcraft and then acquired a larger cruising boat, which I have over a period of years intermittently restored and sailed as far afield as the Florida Keys.

Ernest, on the other hand, like some less-adaptable and long-extinct offshoot branch of early man, has for some reason shown no desire to evolve and change in his boating pursuits. After more than 30 years on the water, he still paddles old, beat-up canoes down southern rivers. And, unbelievably, he still asks me to go with him almost every time! And I often graciously agree, even though he doesn't reciprocate by offering to crew on my sailboat. Even when I found myself abandoned and in need of crew to sail home from the east coast of Florida, I had to rely on my old hometown buddy Jeff Hudson to fill in as first mate. Ernest claims chronic seasickness as his excuse for not answering the call of the sea, a convenient dodge to exclude participation in 90 percent of my adventures. Yet, even though I've recently developed a life-threatening allergy to bee and wasp stings, I'm not exempt from duty as a canoeing partner. Never mind that these southern rivers are literally overhung with dangling hornet's nests, not to mention the streamside thickets that conceal red wasps and yellow jackets.

Like a popular bumper sticker says: "I'd rather be sailing . . . ," but it was too late for that now that I was committed to the river. I had to paddle the 159 miles of the Chickasawhay *and* the 81-mile length of the Pascagoula River before I would see my beloved Gulf of Mexico again.

Now if this had just been any old canoe trip, I would have told Ernest to have at it and would have enjoyed spending two weeks in April anchored off the barrier islands instead. But no, this was far from ordinary. This trip had a higher purpose. This trip would result in a book. Ernest has written lots of books, sixteen, at last count, I think. As I've said, I've known him since he finished the first one. Each time he writes one, he says: "That's it—no more." He claims he has nothing else he wants to write about. Then, within a month, he's hacking away at another one. Watching this for so long has become infectious. I've always liked to read, and even when I was a kid thought I might like to write a book: probably some Tarzan-style fantasy adventure novel. But it was not until after my first long kayak trip, an adventure in the Caribbean, that I ever gave it much serious thought. I wrote an article for *Sea Kayaker* magazine, and that led to subsequent articles. In the meantime, as the years went by, Ernest was churning out books: narratives, children's books, novels, and, finally, guidebooks. While working on his canoeing guide to the rivers and streams of Mississippi, he suggested I compile a similar guidebook to the coastal waters of the state, the area I had come to love after years of sea kayaking and sailing there. To my surprise, my proposal was accepted and a contract for the book was issued. I labored through it for more than two years in my spare time, for the first time understanding the level of commitment all authors must summon to complete an average-sized nonfiction book. It was a lot of work, but at last it was done; then I landed another contract for a book about my Caribbean kayak trip, and now Ernest was offering to let me in on coauthoring a book he had long wanted to write about paddling the entire Pascagoula River system. How could I say no?

It didn't matter that I had already paddled the entire Pascagoula River before, both downstream in a canoe and *upstream* in a sea kayak. Those trips had been over 10 years before and we needed fresh information. Ernest and I had both paddled significant stretches of the Leaf and Chickasawhay rivers as well, but with this trip, his idea was to tie it all together with one epic descent of the entire river system. The resulting book would be a travel narrative. Like him, I worried how we would pull that off—a whole book about a two-week trip? Just days before we left I finished and submitted the manuscript for my second book, a narrative of a trip that stretched over a 13-month period. That manuscript was almost 90,000 words in length, and only covered the high (and exceptionally low) points of the trip. To write a book-length manuscript about this trip, Ernest and I were going to have to take a different approach.

The higher purpose of the trip—that is, the book and the resulting publicity our efforts might bring to raise public awareness of this magnificent resource—won out over my reluctance to pick up the paddle and spend two weeks far from the sea. At least I wouldn't have to paddle with Ernest. Not in the same canoe, or even on the same river (except for the last 4 or 5 days when we would travel the Pascagoula itself after meeting at Merrill). Two previous river trips a few years before had cemented my resolve not to get stuck in the same canoe with him for more than a few days at a time. One was our descent of the Rio Patuca, in Honduras, in 1995, and the other was a trip down the Tennessee-Tombigbee Waterway in 1996.

On the Patuca, I wanted to linger. I wanted to savor the rare experience of paddling through a true tropical rain forest. I wanted to see the wildlife: the monkeys, tapirs, maybe even a jaguar, but that would require sitting still in concealment for hours. I wanted to stop in the Miskito Indian villages we passed, to try and learn something of their culture, to make friends, to trade for souvenirs. I knew I might never pass that way again, and wanted to savor every minute of it. Ernest, on the other hand, wanted to make miles. He wanted to push as hard as possible to the Caribbean coast where the trip would end. So as usual, we got to the end of the river days ahead of schedule, and had nothing to do but wait. All that backbreaking paddling so we could swelter in the torrid heat of Brus Laguna for half a week, swatting flies and longing for a cold beer that was utterly unobtainable.

And on the Tombigbee Waterway, we paddled for days on end across vast open lakes and down man-made canals, into driving rain and headwinds. Big waves threatened to swamp us in Ernest's battered Old Town Canadienne canoe, as we bent to the paddles under our ponchos in the freezing spray and rain. I swore off paddling big water in open canoes and longed for my kayak, where I would be snugly sealed in the cockpit and low enough to slip under the brunt of the wind. Even on my worst day, I could easily outdistance two paddlers in an open canoe in such conditions, and remain comfortable doing it. I'm still baffled by Ernest's resistance to switching to kayaks when situations like that make them the obvious choice.

On this trip, he could canoe however he wanted. I would take my sea kayak, and paddle my own river. This way we would both be happy and we could each travel at our own pace. When it came time to decide which river we each should paddle, it was only logical that I should take the Leaf, for reasons Ernest already derided in chapter 1. It's true that my family, on my father's side, has its roots on Bowie River, a tributary to the Leaf. My father grew up on a working cotton plantation there, and even today 200 acres remain in the family, 80 of which are forested in bottomland hardwoods, through which the upper reach of Bowie River runs cold and swift. Those bottomland forests captured my imagination from my earliest memories. I followed my dad across sloughs and through the dense bay thickets at dawn, an old single-shot 20-gauge cradled in my arms. We would find a big white oak or pin oak stand and sit against a tree in silence until the squirrels that had watched us walk in forgot about our presence and resumed their noisy feeding. But to my impressionable mind, fed on too many jungle movies and comic books, we were the hunted as well as the hunters. I was certain that a black panther or some other imagined beast would appear at any moment, and, walking behind Dad either coming or going from our hunting grounds, I cast frequent glances over my shoulder to see what might be following us.

On the plowed land just up the hill from the Bowie River swamps, arrowheads and other stone tools turned up on a regular basis, and my grandmother told us about how her grandmother remembered when a band of migrating Indians still passed by and made camp near the river every year.

It seemed a natural choice that I should paddle the Leaf and discourse on its tributaries such as the Bowie River. After all, I had lived most of my life within an hour's drive of the river, and had already accumulated many experiences on the river from several previous paddling trips. But Ernest wanted to paddle the Leaf as well, partly because it was shorter—some 135 miles as opposed to 159. Since I would be traveling in the faster boat—a sea kayak—he reasoned that I should have to paddle farther. But hey, is it my fault I prefer a sleeker, swifter craft that makes days of 25, 30, even 40 or more miles not only possible, but relatively easy? In the interest of fair reporting though, since Ernest lives far from both the Leaf and the Chickasawhay, and does not have any deep-rooted connection to either river, I was okay with letting him have the Leaf. That way we would both have to open our minds and our eyes to what we found during the days we were paddling, and neither would have an advantage over the other in terms of the geography.

Ernest mentioned the job we were offered by the Pat Harrison Waterway District, and implied that my decision to paddle the Chickasawhay might somehow have been influenced by monetary considerations. I should take this opportunity to set the record straight on this matter. The Pat Harrison Waterway District had obtained a survey of river obstructions (such as fallen trees) in the Chickasawhay River in 2002. When Chris Bowen, director of the district, learned of our proposed journey and book, he saw an excellent opportunity to have the findings of the 2002 report checked and verified. This would eliminate the expense of sending a team of his own employees out on the river in a motorboat, so he could pay Ernest or me a moderate sum to check this data on our way down the river. The only catch was that collecting the data involved using some relatively simple technological devices, namely, a digital camera and a Global Positioning System (GPS) receiver. Ernest stubbornly refuses to adapt to changing technology (as mentioned before in his choice of boats) so he wanted no part of anything that would involve this high-tech equipment. Using a GPS receiver is everyday practice for sailors, and I had done some surveying work before, so in the name of science and as a public service for taxpayers who support the waterway district, I graciously agreed to change my plans at the last minute and do the Chickasawhay. It was a win-win situation. Ernest would get to do the easy Leaf. The Pat Harrison Waterway District would save some money and get their report verified and updated, and I would get some extra cash to help pay for the expenses of a two-week trip. Unlike Ernest, who has a full-time income as a reporter for a daily newspaper, complete with benefits and paid vacation (not to mention royalties from all those books), I am self-employed and only get paid when I produce work, so this was income that I needed.

But now it looked as if the rain might sabotage my efforts to gather the data. I pondered what to do if it continued for much longer than a day; should I wait it out, conduct the survey, and risk being days late to meet Ernest and Travis? Or would I eventually have to forego doing the work and forfeit the money I would earn in the interest of staying on schedule and getting down the river in time to make our prearranged takeout in Gautier? I sat in the dark and wondered how long it would rain. I had no radio to check the forecast. My last weather report had been direct from Walt Grayson, the veteran ex-weatherman from WLBT Channel 3, who had come down to the Leaf River the day before to shoot some footage of Ernest and Travis launching for his "Look Around Mississippi" program. Walt said it was going to rain, and it was. The rain was supposed to be heavy Saturday night, Sunday, Sunday night, and end sometime Monday. If that was accurate, I would only have 7 ½ days to paddle 159 miles.

With the flashlight, I pored over the 2002 Pat Harrison Waterway District report. There were over 160 charted obstructions in the report. I would have to stop at each one, take a digital photograph, and write notes in my field book describing whether or not the snag corresponded to the one in the report. In addition, I was supposed to photograph, describe, and record GPS coordinates on any new snags I encountered. Plus I had to photograph all the bridge abutments and boat ramps I encountered on the entire length of the river. There would be no more sleep tonight. I had too much to worry about, and the thunder was keeping me awake anyway.

The hand-built Arctic Tern sea kayak pulled along side the bank on the Chickasawhay River.

9

Daylight came with no letup in the rain, so I sat up in the tent and ate a cereal bar and some dried fruit, lamenting the fact that I had left my stove and coffee pot in the kayak at the foot of the steep bank. It was just as well that it was still raining though. I had a lot to do before I could get moving downstream. I powered up my handheld Garmin GPS receiver, and began punching in the waypoint data for the first 40 or so obstructions described in the Pat Harrison Waterway District report.

There had been a moment of panic the afternoon before, when I reached the beginning of the Chickasawhay River after fighting the current upstream from the bridge at Enterprise. As the previous survey of river obstructions conducted in 2002 had begun at the confluence of the Chunky River and Okatibbee Creek, I was also required to begin at this point to conduct my comparison survey. The first waypoint in the report was taken at the confluence, along with a corresponding digital photograph looking upstream into the mouth of each of these waterways that form the Chickasawhay. After struggling against a swift current of shoal water running over a rock and clay bottom, I nosed the kayak into a mud bank where it would stay put while I readied my own digital camera and GPS to mark the official beginning of my 2004 report. When the GPS acquired a position fix after a few seconds' delay, I was dismayed to see that the numbers did not jibe. According to the report, the coordinates for the confluence were: 88.820714 West, and 32.18251 North. The numbers on my screen were: 88.49.252 West, and 32.10.942 North. How could this be? There was no question as to whether I was in the correct location. The photograph in the report looked exactly like what I was seeing. I looked at the list of waypoints on the report again. At the top of the page was a note stating that all waypoints were given in *decimal coordinates*!

I couldn't believe it. Why had I not seen that note before I left? I had never used decimal coordinates. That was surveying stuff. My use for GPS in the past had been strictly for navigation, and in navigation, the standard way of measuring latitude and longitude coordinates is in degrees, minutes, and seconds. It was the default setting for my GPS receiver, and I had never given it a second thought. I knew that the man who conducted the previous report had used an expensive Trimble GPS unit designed for surveying. It was obviously more accurate than the low-end consumer unit I carried, but I had failed to consider that my receiver might not even be capable of working in the same units of measure. Had I given up the chance to paddle the Leaf River for nothing? Had I spent hundreds of dollars on a new digital camera and extra memory cards for nothing? If I couldn't check in to the waypoints on the 2002 survey, there was no way I could complete the job. I didn't even have a calculator to convert degrees, minutes, and seconds to decimals, and even if I did, it would take too long and make it incredibly difficult to find over 150 waypoints as I paddled 160 miles in less than 9 full days.

The problem could not be solved as I sat there in the kayak with the sun rapidly sinking into the tree line and no decent place in sight to camp. I went ahead and wrote down the on-screen coordinates I had in my field book and took my digital picture of the confluence. The next two photos in the 2002 report showed deadfall trees that I did not pass on my way upstream from Enterprise. There were, however, two newly fallen live river birches that formed what I deemed to be a significant obstruction to the channel, so I photographed them and wrote down the coordinates. By this time, it was getting darker, and I knew that a good open campsite on a sandbar was out of the question. I was afraid to paddle much farther down without knowing if there was a way to read the decimal coordinates on my GPS, so, not far below the confluence, I landed on a tiny sandbar just wide enough to rest my kayak on. It was a good 20 feet up a steeply sloped mudbank to level ground. I tied the bow painter of the kayak to a protruding root and scrambled up to take a look. There was a place to pitch a tent atop the bank, but first I would have to take out some red bay saplings with the machete to make a space. I knew thunderstorms were in the forecast, so I cast a nervous glance overhead into the tall hardwood trees above this questionable campsite, looking for large dead branches that could crash to earth in a high wind. Frontier travelers often called such hazards "widow-makers" and no doubt in the history of woodland travel many have met their end as a result of not paying attention to what was over their heads. I don't like camping under such trees in bad weather if it can be avoided, but this time it could not. There was solid forest everywhere. I pitched the tent under the trees with resignation, taking some comfort in the fact that I was divorced, which could be a decided advantage in this situation.

Ernest scoffs at the idea of carrying cell phones on a canoe trip, and I would have too, not so long ago. I never owned one until just two years before, long after constant communication was a way of life for most Americans. But when I did acquire one, I found that rather than becoming a burden, the phone gave me more freedom, as I could go where I pleased and not have to worry about missing important calls that might mean work for a self-employed carpenter.

Regardless of our personal tastes or distastes for technology, Ernest and I both carried cell phones (his on loan) on this trip so we could arrange our rendezvous and call in to Paul Ott's *Listen to the Eagle* radio show, as Ernest already mentioned. I didn't expect to have service in most of the remote areas through which the Chickasawhay runs, but since the beginning of the river is just a few miles away, as the crow flies, from I-59, when I turned on my phone, I discovered that I had a strong signal.

It was imperative that I find a solution to my GPS problem. I had failed to bring the owner's manual, as I thought I knew everything about it I would need to know, so I had to seek outside help. I called my brother, Jeff, a professional construction layout surveyor, and he told me how to go to the set-up menu in my receiver and change the map datum units to decimal. With that crisis solved, I could now enter waypoints while I waited out the rain. The waypoint coordinates were stored in the unit's memory, each labeled with the same name assigned in the 2002 report. All I had to do as I paddled downstream was set the "Go To" feature of the GPS on the next waypoint en route, as shown on my detailed topo maps, and paddle until the numbers showed I had reached that destination. Then I would get out the photo from the previous report, see if there was any downed tree, stump, snag, or other obstruction that resembled the one in the 2002 photo, and take a corresponding photo from the same perspective. Before moving on, I would also have to describe the obstruction in my field book.

By the time I had the first 40 or so sets of coordinates plugged into memory, the rain slacked off enough to allow me to brew a tall cup of strong black coffee. There's nothing like hot coffee and camping. It's unfathomable to me that Ernest could give it up for tea. I didn't even drink the stuff until I started camping with him. Now he's quit and gone on some "purification" kick with expensive, exotic teas. He did that with cigars too, after introducing me to the pleasures of puffing them around a campfire almost 20 years ago. But despite his abstinence, I had a good supply of both on this trip—three varieties of coffee, and enough Irish cream-flavored cigars to allow one after lunch and one after dinner every day for two weeks.

By 11:00 a.m., the rain had mostly stopped, so I broke camp with plans to paddle at least as far as the first sandbar, where I could hope to find much better camping conditions for Sunday night. An added bonus of doing the Pat Harrison survey was the supply of large-scale topo maps and even aerial photographs that were included in the report. With the aerials, I could see practically every sandbar on the river, as well as which areas were wooded, which were pasture, and where roads and other signs of humans were present. A series of sandbars showed up in the photographs just a few miles below the boat ramp at Stonewall, the next point of road access, 4 miles below Enterprise. It would definitely be worth moving at least that far downriver, just to have a decent place to spend the night, even if it did mean a short day.

In the process of getting my gear down the steep embankment upon which I was camped, I managed to get mud smeared on just about every piece of equipment and clothing I had. The tent was soaking wet, but there was no alternative to rolling it up, wringing out the excess water, and shoving it far into the bow compartment of the kayak. Packing a 23-inch-wide, 17-foot kayak with 2 weeks' supply of food and camping gear, plus photography equipment, is an intricate art in the best of conditions. When Eddie McCalip dropped me off under the Enterprise Bridge the day before, it was already late and I had little time to meticulously organize my gear. I knew it would be a close fit to get everything I needed inside the boat, so I had test-packed it a week before at home so I would know exactly what I could and could not bring. What I didn't do was test it in the water for load-carrying capacity.

My workhorse long-distance traveling kayak is a Necky Tesla, built of Kevlar and the veteran of both my Caribbean and Canada-to-Vicksburg voyages. The Necky is also 17 feet, but has a 25-inch beam amidships and lots of interior volume due to its deep bilges and full bow and stern.

It would easily swallow up the gear and food needed for this trip, but it was not the boat I wanted to paddle.

Despite the proven reliability and still-good condition of my old kayak, I wanted to paddle a boat that would inspire me and that could capture my imagination when I was just looking at it beached on a river sandbar. I wanted to paddle a conversation piece that would turn the heads of other river travelers I might happen to meet and break the ice for conversation that might lead to new knowledge of the river. The kayak that could do that was a sleek, handmade mahogany vessel that I had built myself: a model called the Arctic Tern by its designer, John Lockwood. A wooden boatbuilding kick I had gone through for several previous years led to a frenzy of turning out small boats, both for my own use and for paying customers. This particular kayak had been built on commission for my nephew, Brian Nobles, who fell in love with its striking lines and toyed with the idea of embarking on long journeys like his crazy uncle. Brian's kayaking career was short-lived, however, as he was soon hard aground on the shoals of marriage and fatherhood. The kayak that once served as an expensive and impressive coffee table, gleaming beneath 10 coats of varnish in his bachelor's apartment, was banished to a corner of his suburban backyard. There it had begun the slow process of returning to nature when I rescued it by buying it back for a fraction of the sum Brian had originally paid me to build it. With a minor sprucing up, the Arctic Tern was as good as new, and I longed to take it on its first real adventure.

Patterned after the hard-chined sealskin kayaks of the indigenous people of Greenland, the Arctic Tern kayak requires no rudder to track straight, as do many modern sea kayaks. It is also capable of quick maneuvers and graceful leaned turns when handled by an experienced paddler. Built of exotic African ocume plywood, just 4 millimeters thick but sheathed in a protective layer of transparent 6-ounce fiberglass laminated with epoxy resin, the Arctic Tern is as strong as any production sea kayak, yet at only 39 pounds is lighter in weight than most. And I already knew from miles of day trips in it that it handled and paddled like a dream. The simplicity of no rudder, with its connecting cables that could get hung up in overhanging branches on a twisting river, was another appeal of using this kayak for this particular trip. It was an elemental boat, not that far removed from the kayaks of the far north that inspired it, and I loved it for its utter simplicity.

All boats are compromises, however, and what I would compromise by paddling this narrow kayak with its missile-like bow and stern was interior space and load-carrying ability.

That first day, while Eddie McCalip set up his video camera and asked me questions about my kayak and the trip for the *Listen to the Eagle* show, I crammed all the supplies and gear I had brought to the river in two huge duffel bags into the kayak. The previous two days at home had been spent repackaging food items and most everything else I carried into waterproof, one-gallon Ziploc freezer bags. Larger items were sealed in purpose-made waterproof dry bags. When the boat was stuffed to capacity, there were still many items that had to be lashed on deck, such as my Therm-a-Rest mattress, spare paddle, PFD, chart case, tripod, and, in a new waterproof gear bag, my SLR camera with two zoom lenses. I would never load excess gear on the deck of a sea kayak for ocean paddling; it raises the center of gravity too high and inhibits rolling ability. But after years of rough water experience, I tend to get casual about preparations for gentle rivers like the Chickasawhay.

I shoved the kayak off the steep sandbar under the bridge and then stuffed the final dry bag, the one containing rain gear, snacks, sunscreen, and other essentials needed close at hand, into the cockpit between my knees. With all this gear and my 200 pounds, the kayak sank dangerously low in the water and felt tippy, as if the slightest movement to one side could roll it right over. I was thankful I was not launching off some beach into open water with a long crossing ahead of me. There was nothing to do but go on. I knew that in a few days, I would literally eat my way into the boat, gaining more space and carrying less weight as food supplies were consumed. I pointed the bow upstream and dug the paddle blades in against the swift current. Eddie had the camera on me until I was almost out of sight; then I was alone as the Enterprise Bridge slipped from view around a bend and the forest closed in around me.

As I was going upstream to the start point, I knew from my topo maps that the river split around a large island, forming two channels just below the confluence of the Chunky River and Okatibbee Creek. There were snags charted in the Pat Harrison report on the east fork, so that was the one I had to take in order to reach the confluence. Though it was only about a half mile to the beginning of the river from the bridge, for most of the way the river was swift, including a section of rocky rapids that forced me to get out and wade, towing my boat upstream behind me. In the sections where I could make headway against the current with the paddle, I had to be careful to keep my bow pointed into the flow, because I felt certain that in its overloaded condition the kayak would be easily capsized if it became beam-on to some of the swifter sections.

Now that I was reloading the kayak on this rainy Sunday to begin the long journey downriver, I wanted to pack carefully so that I could take advantage of every cubic inch and get some of the gear I had on deck down to a lower point in the boat. But on the slippery mud bank of that first campsite, organization was difficult at best, and I contented myself with emptying the extra water jugs I carried and thus shedding a few pounds off my load. I had Polar Pure iodine crystals to purify drinking water, so there was no need to carry extra water beyond that of my two one-quart drinking bottles.

The return trip downstream to the Enterprise Bridge took only a few minutes, and soon I dropped it astern after stopping briefly beneath it to record new coordinates now that I had my GPS configured to read the position in decimals. Below Enterprise the river slowed, meandering past high bluffs of clay that dripped and gurgled with runoff from last night's rain. Gray overcast promised more rain, and the woods that closed in around the river were unusually quiet, the leaves of the hardwoods glistening wet and impossibly green. To me the best time to be in the woods is after a heavy rain. That's the one time when it's possible to walk across fallen leaves in total silence, and the best time to catch a glimpse of a deer, turkey, or other wary wildlife as the forest comes back to life and its inhabitants resume the never-ending quest for food.

There were only three significant obstructions in the river between Enterprise and Stonewall. All were new, and all were fallen trees with green leaves still on them. I recorded the coordinates and took photos—no big deal if all the snags I encountered were that easy, but I knew they would not be.

As I approached the bridge at Stonewall, I saw that the ramp just downstream on the east bank was occupied by three fishermen: a black man in his 20s, an extremely thin white man with long, scraggly hair who seemed to be about the same age, and a boy who looked about 13. The two men each had a rod and reel in one hand and a beer can in the other, and it was evident that they had been there awhile by the accumulation of empties scattered about their feet. They were standing right in the way of the photo I was required to take of the boat ramp. Normally when I pass other boaters or bystanders on a river trip, I just politely wave or speak and continue on my way. I could see this trip was going to be different. I had to stop and take photos and coordinates of every bridge crossing and boat ramp along the way, and these were the very places where I would likely encounter people like this hanging out by the river, especially on weekends.

I paddled close enough to be in range of easy conversation without getting too close to their fishing lines. I would have to explain my need to photograph the ramp, but it was pointless to say I was doing this trip for the sake of adventure, or even for the writing of a book. That would be incomprehensible. My boat, so small and long and narrow, and above all, lacking a motor, was foreign enough without added complications. I told the fishermen that I was a surveyor, floating the river to get information on the bridges and boat ramps. This seemed acceptable to them, and the two grown men stepped aside so I could take the photo, but the teenager didn't seem to mind being in it.

"You work out here like this all the time?" the long-haired man asked.

"Well, not just on this river."

"Where you taking out tonight?"

"I'm not. I'll just keep going until it gets too late to see, then I'll camp out on a sandbar."

They were all amazed at this.

"I hope they're paying you good!"

"It's a job . . . and I like being out here," I replied. I answered their further questions about my boat: How hard was it to build? How much did it cost? And about where I was going: All the way to the coast? How long would that take?

"I didn't know this river went to the coast," the black man said.

They all do, I wanted to say, but I just told them it went to the Pascagoula River, then from there to the coast, and then I told them how many miles it was, and how long I thought it would take to get there.

"But I could make it in less than 2 weeks, probably 10 days if I wanted to," I said.

The long-haired man turned to his buddy and grinned. "That's what he's gone do. He'll bust butt and get down there ahead a time so he can lay up drunk at them casinos while his boss thinks he's working. That's what I would do!" I laughed with them as if this was exactly what I planned to do, and then waved goodbye to paddle on. They wished me luck and a safe journey and watched until I disappeared around the first bend below Stonewall.

Dawn on the Chickasawhay River above Shubuta.

10

The fishermen at Stonewall Ramp were the only people I would see for the next 3 days. Although the Chickasawhay River is certainly used by fishermen in motorboats and even the occasional canoe, it was afternoon on Easter Sunday and still overcast and gray with the promise of more showers soon. I knew that after the weekend river use would be down anyway, and the bad weather made it even more unlikely I would see anyone out on the water.

That Sunday afternoon I paddled silently past a few riverside houses or weekend cabins, but no one was outside to see me go by. The river was deserted, and in a way I enjoyed the feeling that I was threading my way down its twisting course in secrecy, with no need to explain myself. I had been warned by the Pat Harrison Waterway District that certain landowners in Clark County were the primary reason all this money was being spent analyzing the feasibility of removing snags from the river. Apparently, some of them with a lot of riverside property thought there should be a wide-open, obstruction-free channel that would allow them to speed up and down the river in a powerful motorboat. I had been told that it would be best if no one in this area knew I was on the river rechecking it for snags, so I didn't plan to go down the river blabbing about what I was doing. That was the reason I didn't mention snags to the guys at the Stonewall Ramp. I just gave them the impression I was surveying the bridges and boat ramps for the state highway department. Apparently these vocal Clark County landowners advocated a total removal of all natural river obstructions, which, according to the 2002 report, would amount to a staggering cost. It would also severely alter the natural characteristics of the stream, and remove a lot of valuable underwater habitat for fish and other aquatic life that depend on such cover for spawning and hiding from larger predators. And based on what I had seen in just the first few river miles, new trees would always be falling into the river, requiring constant effort and expense to keep the channel clear. It was already obvious to me that my catalogue of new obstructions would only serve to drive home the point that such a massive de-snagging effort was futile. But I didn't want to have to explain this to some irate landowner who might prefer that I didn't conduct the comparison at all.

A kayak is small, fast, and above all silent, making it the ultimate watercraft in which to conduct such a surreptitious survey undetected by the local residents. As always when traveling rivers or coastlines by kayak, I planned to keep a low profile anyway, especially each evening when it came time to camp, as many landowners are intolerant of even low-impact campers as a result of the abuses of the other variety. I thought of all the places I had camped as I paddled down the lonely Chickasawhay and hoped that as in the past, I would avoid trouble with such people as I had always managed to do in my other travels. In wilder country, this is seldom a problem. Many of my trips traversed big wilderness areas that were public domain in the form of parks, national forests, or other government-owned land. Mississippi's hundreds of miles of floatable waterways course through mostly private lands, however, requiring some discretion on the part of river travelers needing a place to spend the night.

Attitudes about private land in Mississippi have changed much just in my lifetime. When I was a small boy learning how to hunt and handle a shotgun by tagging along in my father's footsteps, the bobwhite quail was one of the most sought-after game species in the state. With Charley, my brother's English setter, or Babs and Tick, my dad's liver-and-white pointers, we would take off across fields of broom sage, soybeans, or cotton seeking the elusive coveys that could only be found with the aid of a good gun dog. Dad usually bagged 2 or 3 birds on the covey rise with his slick-handling Remington 20-gauge pump, and then we took off with the dogs in search of the singles. This could take most of a day, and entail miles of busting through briar patches, working around cutovers, and crossing more fields. We thought nothing of crossing property boundaries, and no one else did either. It was accepted practice to hunt most anywhere you pleased, so long as you closed gates behind you, avoided damaging barbed-wire fences, and didn't trample crops or endanger livestock with misplaced shots.

Then in the span of just a few years, deer hunting took over as the preferred form of outdoor recreation and everything changed. Although deer were rarely seen in many parts of the state in the 1960s, by the time I graduated from high school in 1981 they were widespread and abundant, and deer hunting was a major part of rural culture. Unlike quail hunters and small-game hunters who went after squirrels and rabbits, deer hunters affected the lands they hunted on in ways that angered and alienated landowners. Using dogs, high-powered rifles, and all-terrain vehicles (ATVs), deer hunters were perceived as dangerous and destructive. In addition, deer hunting was more competitive than small-game hunting, and those landowners who did have large chunks of deer-rich woods became possessive of game they perceived as theirs. As a result of all these changes, posted signs went up everywhere and the tradition of hunting the old way, by roaming for miles across whatever land upon which the quarry could be found, came to an end.

In addition to hunting, increased river use due to the ease of access offered by outboard motors, and in some places, 4-wheeler ATVs, has led to more restrictions on riverbank access. Technically, the landowners with property adjacent to the river also own the streambed to the middle of the channel, including sandbars both submerged and above water level. Such landowners can refuse access to boaters wishing to stop or land on their property; they cannot however, prevent anyone from using the river if it is deemed a navigable waterway (including navigable only by canoes and other small boats). Paddlers traveling the creeks and rivers of Mississippi seldom have to worry about restrictions on bank access, however, as there are only a few problem areas on popular streams where large numbers of drunk and rowdy weekend floaters have had run-ins with the landowners.

In general, as Ernest and I have discovered over many years and hundreds of miles of canoeing, the owners of stream-side property are seldom, if ever, present, and traveling these streams, one gets the impression that no one at all owns the vast tracts of woods that surround them. This is because the owners live elsewhere, seldom building near the banks of the river and in many cases never even venturing into the deep forests and bottomlands they own. As the ancient Taoist sage Lao Tzu said: "Water flows in the places men reject."

Avoiding confrontations with landowners on a camping trip is mostly just a matter of common sense. As Ernest already mentioned, all a river traveler has to do if asked to leave a sandbar is simply paddle on around the bend to find an unclaimed one. In more populated areas I've traveled through, however, especially coastal areas where camping was definitely illegal or at least unlikely to be tolerated, I've had to resort to dragging my kayak into the concealment of the bushes and camping out of sight of the waterway in the forest. In many such "commando" camps, I've quietly passed an evening undetected by swarms of passing motorboats and even marine patrol officers. Slipping back into the water before dawn, I would paddle through cities and towns in the daylight, mistaken by all who saw me for a casual day paddler out for an afternoon jaunt.

I did not at all expect to have to resort to such extreme measures on the Chickasawhay River or even the Pascagoula. I planned to camp out in the open on the sandbars, only taking care not to spend the night on one accessible to vehicles or ATVs if at all possible.

The Chickasawhay River had flooded far out of its banks earlier this year. I saw evidence of that when I first launched at Enterprise, in the amount of debris that was still hanging from the branches of riverside bushes and trees. A fine coating of dried river mud covered the tree trunks, branches, and leaves to a level of at least 20 feet above the current river stage. I tried to imagine the raging torrent the Chickasawhay must have been in such a flood. This was part of natural ebb and flow of a wild river, and the entire Pascagoula River basin still benefits from these periodic inundations that spread nutrients across the floodplain and sweep debris downriver to the Gulf. Now the river had subsided to a level more typical of a dry late July or August. Even so, I had no problem finding a deep-water channel, and not once did the hull of my kayak touch bottom on these upper reaches of the river. This was not typical of a Mississippi stream, especially so high up in its run. Normally, dragging canoes over sandbars and gravel shoals is to be expected in the daily routine of paddling downstream.

Even with the deadfall trees I began to encounter more frequently, both new ones and old ones documented in the 2002 report, I found no obstacles I could not easily maneuver around. A small johnboat with a modest outboard motor would have no difficulty either.

I reached the sharp horseshoe bends I had seen on the aerial photographs, where slivers of white against the infrared background of forest indicated sandbars. There was indeed sand in these bends, and the second one offered a high and level spot just big enough for my tent. Clean sand sloped gently down to the water's edge, offering an inviting place to land and unload my boat. What a contrast to the camp I had left this morning! I lost no time digging out my tent and setting it up with half a dozen long stakes to secure it in the deep sand. Rain was coming soon, and high winds and more thunder and lightning were likely to accompany it. With the tent pitched and my bedroll laid out inside, along with everything I needed to prepare dinner, I then went back to the river's edge to move my kayak to higher ground. I learned the hard way not to trust rivers during the night. They can rise to unimagined levels in just a few hours, even when conditions seem impossibly benign.

I spent over a year of evenings and weekends building my first handcrafted wooden boat—a 19-foot, decked canoe I called *Seldom Seen*. Just a few months after this woodstrip canoe was completed, my buddy Jeff Hudson and I decided to take it on a week-long trip down the Leaf River, starting in Petal and continuing to the Pascagoula and on to the coast—the same route Ernest and Travis were now emulating in their pathetic open canoes of fiberglass and plastic. A few miles below Hattiesburg, Jeff and I spent our first evening in camp in lively conversation, staying up until almost midnight thanks to our other companion, one Senor Jose Cuervo, who came along in a 1.5 liter bottle.

We had launched that day on a sunny afternoon, and stars shimmered out of a clear sky as we talked that night. I glanced at the canoe before we finally turned in to our tents. It was several feet from the water's edge, on the mostly flat sandbar. I didn't give it a second thought. I was awakened from a deep sleep the next morning at dawn by Jeff's shouts. Knowing Jeff is ever the practical joker who can never be taken seriously, I rolled back over and ignored him when he yelled, "Scott! The boat's gone!"

He persisted, which was to be expected of him.

"Scott . . . I'm serious, the boat's gone! Scott! Wake up! The boat's gone!"

Yeah, right. You'll have to do better than that, buddy.

"Scott, I'm not kidding . . ."

"Gone where?" I answered, sleepily.

"The water rose. It's right outside the tents. When I went out this morning, I took two steps and I was splashing!"

Now he had my attention. He was so insistent I would at least have to humor him by looking. I unzipped my tent door. A wall of gray fog obscured everything farther than 15 feet away. Incredibly, the edge of the river was about 3 feet from my tent door. Jeff was standing in it. I bolted from the tent and took off running downstream along the water's edge, trying to see into the soup of fog that hid everything from view. Jeff was right. *Seldom Seen* was gone. The Leaf River had risen several feet in the span of a few hours, and there was no telling how far downstream my labor of love had drifted.

It could have been much worse. As it turned out, we hiked out to a road less than a mile away, then hitchhiked to a store with a pay phone, where we called my nephew in Hattiesburg to come get us. Brian drove us to New Augusta, the next town downriver, and there I found a game warden willing to take me upstream in his patrol boat. We found *Seldom Seen* hung up in some bushes less than a mile below our camp. The boat was saved and the trip was salvaged with nothing but an embarrassing delay that cost us less than a day.

Since that happened, I've been determined not to undergo a repeat performance. I *always* tie my boat to something solid when camping for the night, no matter where, when, how long, or what conditions I expect. Traveling solo, I am doubly cautious. I would be camping on relatively low sandbars most every night on the Chickasawhay. I pulled the kayak all the way up near my tent and tied the bow painter to a stump protruding from the sand. A light rain started falling just as I was getting ready to set the stove up and cook some rice for dinner. I retreated to the tent to wait it out. At least there were no mosquitoes, and the rain wasn't hard enough to force me to close the tent door. I sat cross-legged in the opening and stared out at the river. With a couple more hours of daylight left, I had time to study my topo maps and aerial photographs of the route ahead. At the rate I was going, it seemed impossible that I could check into all the waypoints on the report and still make the minimum daily mileage I would need to get to Merrill on time. I punched in the coordinates for the next 20 or so obstructions and tried to calculate where I might be by nightfall Monday.

I hoped I would have cell phone service, wherever it was. Ernest and I were both supposed to call in to Paul Ott's live radio show, *Listen to the*

Eagle, at around 6 p.m. on Monday. At least I had service tonight. Apparently I was still fairly close to I-59. The tower indicator on my screen showed 3 bars. I called my girlfriend, Michelle. It was wonderful to have a means of communication, no matter what Ernest's opinion of people who call their wives, etc., while on camping trips. (I was to find out later, from Travis, that Ernest used his borrowed cell phone to call his wife, Angelyn, practically every night! Hypocrisy!) I could only talk a few minutes anyway, because of the need to conserve my battery, so what could that hurt? I had the other 23 hours and 50-odd minutes of the day out here to commune with nature and get away from it all. And besides, when I really want to get away, I head out on the Gulf in my sailboat. Beyond the horizon I'm really out of touch, at least with the limited technology I have on board.

The call had to end all too soon, but I felt better having shared a bit of what I was experiencing with Michelle. The only downside was that she had bad news as far as the weather was concerned. It was pouring in the Jackson area, and more rain was forecast—not just for Monday morning, as I had first heard—but for all of Monday, Monday night, and into Tuesday. She had been watching the Weather Channel before I called. Despite the fact that I carried a phone, I did not bring a radio along. On coastal kayaking trips, of course, I always carry a handheld VHF transceiver, which is capable of receiving NOAA weather forecasts. But here on the river, with no need to call other boat traffic and no expectations of any weather that would pose a threat, I had no use for such a radio. On longer trips I might also carry a Walkman with a built-in AM/FM receiver and some favorite CDs. Ernest would again scoff, but then Ernest has never spent more than a week traveling solo. Music can be nice on a long trip, where bad weather can sometimes force a paddler to lay over for days on end in the tent with little to do. On this trip I could do without it though. There was simply no room in the kayak for such luxuries. I was packed to the utter limits of my storage capacity, and with the Pat Harrison documents, my field book, journal, and a couple of nature guidebooks, I didn't even have room to bring a novel or two for reading material. Oh well, little matter. There would scarcely be time to read anyway, and I figured that time I would spend reading might be better spent writing, or taking pictures.

Before the trip, I perused listings on eBay searching for a larger zoom for my Nikon SLR camera than the 28–90 millimeter lens that came with it.

I found a deal on a 70–300 zoom, and bought it thinking it might help me get some closer shots of any wildlife I might encounter, without being so powerful as to produce blurry images when handheld in a moving kayak. One never knows what to expect traveling solo on a deep-woods river like the Chickasawhay. I've missed more incredible photo opportunities than I care to remember. A whitetail buck swimming across the river in front of me . . . a gray fox trotting along the bank just yards away . . . a barred owl sitting on an overhanging branch in broad daylight . . . a monster alligator in the Pascagoula marshes that had to be about as big as they get . . . the list goes on and on. This time I wanted to be prepared. The camera and both lenses were carried close at hand in a river drybag strapped to the foredeck of my kayak. Even with the drybag, this was a risky way to carry expensive gear, but like a gun, a camera is only worth having if you can get to it when you need it.

During the proposal phase of our Pascagoula book, there was some talk of the publisher paying a top-notch regional wildlife photographer like Steven Kirkpatrick or Joe Mac Hudspeth to shoot the photographs for our book. Then it could be published in the full-color coffee table book format, with Ernest and me writing the text. This was an expensive proposition, and one that got axed because of budget constraints, so now it was up to Ernest and me to provide passable photographs to illustrate our book. Neither of us claims to be a photographer, but we both take pictures to support our writing. I knew that in all probability, whatever photographs were published in the book would likely be printed in black and white, but still, I was hopeful that two weeks on such a beautiful river would yield at least a few decent photos.

The rain let up for a little while on Sunday evening just before dark, allowing me to cook a pot of yellow rice, to which I added a can of Trappey's black-eyed peas seasoned with jalepeno peppers. Rice and pasta are my staples on a river or sailing trip; either can be supplemented with an endless variety of canned foods to make a complete meal, and I never get tired of eating them. It seems that the question "What do you eat when you're out there?" is raised more than any other by noncampers asking about trips like this. But the truth is, my diet in the wilds is little different from what it is back home. I always cook a hot meal for dinner, and usually for breakfast as well—on this trip breakfast would alternate each day between oatmeal and pancakes. Lunches were less planned: cheese and crackers, canned fish, beef jerky, nuts, dried fruit, and, for chocolate dessert that doesn't melt in the kayak, M&M's candy.

The rain returned after I went to bed, and I fell asleep to the steady sound of it pelting my tent. There was no thunder or lightning though, so I slept through the night, waking at first light to see that it was still drizzling and hopelessly gray. I could only hope that the rain would remain light enough to enable me to take pictures. I had to make some miles today, and the next 20 miles or so contained almost half the obstructions in the 2002 report.

Camping on a low sandbar that would be submerged during times of high water.

11

Monday morning dawned gray and drizzly, dashing my hopes the weather front would push rapidly through the area and leave warm sunny skies. That's the problem with planning trips like this based on the calendar. Naturally, there had been weeks of perfect spring weather during the two months preceding our trip. I had been working outdoors building a dock and seawall at a waterfront home on the Ross Barnett Reservoir, and for days there wasn't a cloud in the sky. Now winter-like weather had returned coincidentally on the very week we set for the beginning of our trip. Once again, Ernest could be blamed for this unhappy circumstance. Being self-employed, I could have begun the trip at any time to take advantage of fine weather, but Ernest, on the other hand, holds down one of those 9 to 5 jobs and had to put in a request for his vacation time months in advance. He figured mid-April would be the optimum time of year to be out on the river, and it normally is. Today seemed anything but optimum. As I had done the day before, I wrung the water out of my tent and stuffed it in a soggy roll into the bow of the kayak. The drizzle had quit by the time I had everything loaded and ready to travel, but the river was gloomy and depressing under a leaden gray overcast.

It was at least 10 more miles from where I had camped to Quitman, and this section of the river between Shubuta and Quitman contained more than 42 obstructions charted in the 2002 report. It was going to be a long day. The river here was often narrow, with numerous horseshoe bends, practically all of which had an old or new obstruction in one form or another. Even so, I did not find even one place that required me to get out of the kayak to get through, despite several fallen live trees that stretched bank to bank. There was always some opening just wide enough to thread my way through in the narrow kayak.

The sections of the river with these sharp bends tended to have low banks and lots of sandbars. In one such area, the river had apparently changed course since 2002, cutting through a low bank and leaving a couple of snags charted in the old report high and dry. My preset waypoints for those snags did not appear as I paddled downstream; then suddenly the GPS indicated that they were somewhere behind me to the north. It was a confusing area to chart, and many newly fallen trees added to the mess.

In this stretch of river I found myself constantly stopping to take digital pictures and record descriptions of snags. This left little time for anything else. There were one or two obstructions at practically every bend. The routine went something like this after the recording of each one:

Scroll down on the GPS waypoint menu to the next obstruction waypoint I had entered earlier. The waypoints were assigned numbers from the 2002 report, such as D2-B3, D2-B4, D2-B5, and so on. Enter the selected waypoint number and hit the "Go To" button. The GPS would then show the straight-line distance to the waypoint, in this congested area usually less than a quarter of a mile. Stuff the precious report into my waterproof deck bag. Turn off the camera to save batteries, and stuff it into its own waterproof dry bag in the cockpit. Start paddling while watching the numbers for distance change on the GPS screen. Slow down when the units change from tenths of a mile to hundreds of feet. Get out the report and look at the photo of the charted snag, comparing landmark trees, etc., in the photo with the scenery unfolding before me as the kayak is carried downriver in the current. Frantically get out the camera and wait for it to power up once the obstruction area is spotted, all the while being swept closer to the snag by the current. Try to take the photo from the same perspective as the original, so it will be useful for comparison purposes. Check the screen on the camera to make sure photo looks like original, save it to memory in the camera. Check the latitude and longitude coordinates on the GPS screen, making sure they coincide with original coordinates. Get out the field book kept in a Ziploc bag on deck and write down a description of the snag, and whether it is the same or has changed since 2002. Repeat entire process for next waypoint. All this is done while drifting, to minimize wasted time spent paddling back upstream or trying to stop along the bank. Invariably, the worst obstructions were in places where strong currents threatened to sweep me into dead branches or onto submerged logs and stumps.

I realized that it would take all day to reach Quitman at this rate, and Quitman was scarcely 20 miles into the beginning of the 159 miles I had to make by Sunday. The only consolation I had was that by the time I reached Quitman, I would have one third of the charted obstructions behind me and could hope that the river would gradually get wider and deeper, decreasing the likelihood of stretches as choked as this one.

I didn't see another soul on the river or on the bank all day Monday. I reached Quitman late in the afternoon, and by 5:00 p.m. I had found an isolated sandbar a mile or so downstream from the road crossing. I quickly set up camp and broke out my cell phone, as I was supposed to call Paul Ott's radio show, *Listen to the Eagle*, at 6:00 p.m. sharp. I was now too far from Interstate 20 to pick up a signal on my phone's network, but there was a strong signal on analog roam. This was going to cost me seventy cents a minute, but the cost of calling the media for publicity and calling Michelle briefly each day, if possible, had to be factored in to trip expenses.

I got my list of phone numbers out of my journal and dialed the 800 number Eddie McCalip had given me for the *Listen to the Eagle* show. I was sure they would be expecting my call right at 6:00, as well as a call from Ernest, and I was anxious to hear how far he and Travis had gotten in the first 2½ days. But when someone finally answered the phone after more than a half dozen rings, it was some guy answering the live call-in line who didn't know who I was or why I was calling, even when I explained that I was on the banks of the Chickasawhay River. He said he would have to put me on hold. I thought about the roaming charges racking up while I waited, and I could hear the live feed from the show and listen to the other callers. Ernest was not among them. Some woman was talking about hunting, and after a few minutes I could tell the focus of the show that night was on special hunting days for women. There was no mention of anyone paddling the Leaf or Chickasawhay rivers. I hung up and called back, so I could explain to the guy who put me on hold that I was supposed to call in at 6:00 and also to impress upon him that I was on a sandbar in the woods, on roaming, and with a limited amount of battery power for my phone. Hearing this, he said they would call *me* back, so I left the phone on, thinking I would cook dinner while I waited.

Just as I was about to light the stove, another rain shower rolled in, forcing me to retreat to the tent. Dinner would have to wait. I poured myself a half a cup of Captain Morgan Spiced Rum and sat in the front of the tent with the door open and watched it rain. Rum or whiskey can taste awfully good on a trip like this. I almost never drink anything harder than beer or wine back in the civilized world, but out here, something with more punch is needed, if only to conserve space. For this trip, I carried two pints of Captain Morgan, compact and potent. Just enough to have a drink or two each evening in camp, with a little reserve left over for an emergency like snake bite or something. Though I can't blame Ernest for introducing me to alcoholic beverages, as he did cigars, I can credit him for his wisdom in bringing something to drink on wilderness trips. On our second trip together, a week-long canoe foray into the mangrove swamps of Everglades National Park, his foresight in bringing a fifth of Myers Jamaican Rum saved us a whole lot of misery. The first night out we found ourselves in insect hell at a campsite called Lopez. Almost as soon as we pulled our canoes up onto the bank to stop there for the night, swarms of mosquitoes engulfed us, so thick that when they landed on an exposed body part such as an arm or hand, you could barely see the skin under them. We pitched the tent and dove inside, with nothing between us and the buzzing horde of bloodthirsty bugs but the thin screens of the tent. The rum was our only comfort, and we passed the bottle back and forth until we could at last fall asleep despite the din of whirring wings outside.

Thinking back to that experience, I realized what a blessing it was to be out on the river now with the tent door tied back. I had not been bothered by mosquitoes at all since the first night on the bluff upstream of Enterprise. Compared to so many other places I've camped, such as the barrier island beaches on the Gulf, the north woods of Canada, and the jungles of Central America, not to mention the Everglades, the rivers and streams of Mississippi are usually relatively benign when it comes to insect problems.

An hour after my second attempt to call in to the radio show, my phone still had not rung. I began to doubt they were going to call me back, and I couldn't waste the battery by leaving the phone on. I called Michelle instead and she said she had just tuned in to the show at 7:00 and didn't know if Ernest had called in or not. She updated me on the weather, saying it was supposed to turn cold overnight and colder still tomorrow. That was good news to me. It meant that the rain at the leading edge of the front was moving on through the area. I turned off the phone after our short conversation and put it back into its waterproof bag.

It rained most of the night and, true to the forecast, did turn colder. I had to put on long pants, boots, and a sweater while I made a big batch of pancakes for breakfast. Leaving camp that morning, I faced one of the real disadvantages of a kayak, as opposed to a canoe. Instead of being able to step in the boat and shove off in my nice warm clothes and boots, I had to strip down and change into my wet paddling shorts and pull on the cold, wet neoprene boots that were the only footwear I had room for in the cramped cockpit of the kayak. I only had two pair of long pants and a couple of warm shirts, so it was important not to get those wet, because I needed them at night when I camped. That left just swim shorts, a T-shirt, and my lightweight rain parka for wear all day on the river, and it was obvious already that this was not going to be enough today. I had packed for warmer weather, not really expecting a front of this magnitude so late in the year. Though I didn't have a thermometer, I guessed it was at least in the low forties, maybe the upper thirties, and the wind was blowing out of the north in sudden gusts that rattled limbs and rippled the surface of the river.

The river downstream of Quitman took on a different appearance as it ran straighter and deeper through ravines of steep clay and rock walls. Sandbars were nonexistent in these areas, and the river even exuded a mountain-stream smell, probably because of the rock and clay. Even the steepest of the banks were covered in vegetation—sheets of emerald-green moss and mats of hanging ferns dripping with spring water. But if one could for a moment think this stream were in Appalachia, the presence of towering cypress trees with their buttressed trunks and aerial roots were frequent reminders that this was the Deep South. There was no shortage of subject matter for a photographer here, and I took the opportunity in this snag-free area to get out my Nikon often, but the knowledge that I was still less than a third of the way down the Chickasawhay kept me from lingering as long as I would have liked.

I had hoped that the cold and gray of early morning would soon give way to clearing skies and sunshine, but it was not to be. Instead, by 10:00 a.m., it had turned miserably colder, and fierce gusts of north wind whipped through the deep ravines of the river valley and threatened to tear the paddle from my hands. Massive cloudbanks swept across the sky overhead, driven by the north wind, so there was not even a glimpse of sunlight. It was too cold to be on the river with the gear that I had, and if I had not been so far behind schedule, I would have stopped and made camp with a big roaring fire to keep warm. But that was not even an option. I had to put in a full day and try my best to get as close to the boat ramp at Shubuta as possible. The next day, on Wednesday, I was supposed to meet Walt Grayson at the Shubuta ramp so he could do a follow-up for his "Look Around Mississippi" story. He was driving all the way from Jackson, so I didn't want to disappoint him by not showing up. It was 24 miles from Quitman to Shubuta, so I figured if I could endure the cold all day, paddle at least 15 miles and chart the obstructions on the way, the meeting was still feasible.

Since this section of river was fairly open, I made it to the De Soto bridge crossing just before noon. The town of De Soto is named for the Spanish conquistador Hernando de Soto, who supposedly crossed the Chickasawhay River near here in 1541. Soto and his army of gold-hungry Spaniards were the first Europeans to see this part of North America, when they marched overland from Tampa Bay, Florida, to Georgia, the Carolinas, and then across parts of Tennessee, Alabama, and Mississippi to Arkansas, Louisiana, and Texas. But the town's dubious claim to fame that the conquistador crossed the Chickasawhay River is not supported by historians who have studied his route. It is more likely that the river described in the reports of the expedition was the Tombigbee, much farther to the north.

Hernando de Soto is widely regarded as an explorer and is especially noted for his discovery of the Mississippi River. Actually, he was not the first Spaniard to see the river, as Cabeza de Vaca, survivor of an earlier ill-fated expedition, passed the mouth of the river in small barges they built in an attempt to sail from the Florida panhandle to Veracruz, Mexico. The truth about Soto, however, is that he had little if any interest in exploring or finding a passage to the Pacific, as some have claimed. Arriving at the banks of the great river called the Mississippi was not a moment of joyous discovery for the glory of Spain, but rather a major obstacle to his rampaging march. It took his men more than a month to build boats and ferry his army and horses across this river. His motive was to find gold, pure and simple, and he was determined not to let anyone or anything stand in his way. The story of his army's march through the American Southeast is a story of murder, rape, and pillage. He tortured the native people he met to coerce them into telling him where their "cities of gold" could be found, and each group he met kept sending him deeper into the interior in a ruse to get his army out of their land.

No evidence of the town of De Soto is visible from the perspective of a river traveler, other than the high bridge spanning the gap between almost sheer bluffs and a steep ramp that disappears up the bank on the west side. I landed on a tiny sandbank beneath the bridge and got out to open my bow compartment for some snack food and to try and call Walt Grayson. There was no shelter from the cold as I huddled in my wet shorts under the huge bridge pylons in an effort to get out of the wind. With a weak signal and a low battery, I was barely able to make a call. Walt didn't answer, but I was able to leave a message on his voice mail to confirm our appointment at Shubuta. I was sure I could make it now, as it was just 13½ miles downstream and I still had the rest of the afternoon and the next morning to get there. I called Michelle again to ask her to call Walt also, just in case my message didn't get through. Then the phone shut down, the message on the screen saying "low battery." How could this be, I wondered? Normally, I could talk for a couple of hours between charges, and my total phone time on this entire trip had been less than 20 minutes. Then I realized with disgust that it was because the phone was roaming on an analog network. This required a lot more power, and the brief calls I had made had already sucked the battery down. Before the trip I modified a small solar panel so that it could be connected to the 12-volt phone charger designed to be used in a car. I was counting on this as the solution to keeping the phone charged for the duration of a two-week trip, but without sunshine, a solar charging system is useless. I looked up at the cold gray sky and prayed for sunshine. Not so I could charge my phone, but because there was no way I was ever going to warm up in this bone-chilling cold until the sun broke out of that cloud cover. There was no sign that this would happen soon, but wheeling overhead in the overcast sky was a bird that was an icon of the Pascagoula River system—a swallow-tailed kite—rare almost everywhere else but common on the Pascagoula. This was the first one I had seen on the trip, and I had not expected to see one at all so far upriver on the Chickasawhay.

Before leaving De Soto I got out Ernest's guidebook, *Canoeing Mississippi*, which he had written a few years before, to see what he had to say about the Chickasawhay from De Soto to Shubuta. Of course, I had been with him on his exploratory trip down the Chicksawhay back in 1993, but the details in my memory were fuzzy, and I had not been taking notes. At the time, Ernest was not even thinking about writing a comprehensive guidebook to the rivers and streams of Mississippi. That idea came much later. But he and I had paddled a lot of these waterways together, and his weekly newspaper column of articles about our trips became a handy resource when he decided to write the book. I knew there were stretches of rapids on the river, especially in this upper section where the banks were steep and the terrain almost mountainous. But I couldn't remember quite where. In the Pat Harrison report, the previous surveyor had photographed a sizable rapid just below De Soto. Ernest didn't mention it specifically in his book, but considering how cold and miserable I was already, I planned to be especially cautious in the overloaded kayak with all my expensive camera gear on board. If it did turn out to be a dangerous rapid and I survived it, I would be sure and get full documentation in the form of photographs and GPS waypoints, not for the waterway district, but for Ernest's publisher. I wouldn't let him off the hook if there was a serious omission in his guidebook. Revisions would have to be made and attorneys might have to be retained, especially if my beautiful handcrafted boat was damaged in any way.

I sealed myself into the cockpit with the spray skirt for the first time on the trip, and even put on my PFD, something I almost never do while kayaking, except in the most extreme conditions. This isn't because of a lackadaisical attitude about safety, but is simply because of the nature of kayaks as compared to other boats. You really *wear* a kayak more than you ride in it like a canoe or johnboat. The boat itself is your floatation device, and it is hard to get separated from it once basic kayaking proficiency is developed. Wearing a PFD while paddling is a nuisance, as most designs impede paddling and are hot in warm weather. But running river rapids and negotiating breaking surf on a coastline are two circumstances that make PFDs a good idea, even in a kayak. The main reason is that a capsize could result in a blow to the head on some unseen underwater rock, stump, or other obstruction, and without the PFD, an unconscious kayaker could quickly drown. Drowning aside, the last thing I needed on a cold day like this was an unexpected capsize. I paddled away from the bridge and almost immediately began to hear the sound of swift rushing water. I double-checked that the dry bags secured on deck were clipped into the bungee cords so they couldn't get swept away, and drifted slowly toward the sound of roaring water.

The non-poisonous diamondback water snake is commonly seen on riverside logs and branches.

12

The first set of rapids below De Soto turned out to be insignificant. There was a drop in the river, and an area of small standing waves midchannel where most of the current was funneled, but a lack of rocks or other obstacles made the run a breeze. My precautions with the PFD and spray skirt turned out to be a drill, but I left them on even in the calm water beyond, as there was still little hope of sunshine and the blustery north wind seemed to be getting even colder. It felt like January or February, rather than April.

I stopped for lunch on a sandbar an hour's paddle from De Soto and crouched in a grove of willow trees to get out of the wind while I quickly ate a can of fish steaks in mustard sauce with Wheat Thins crackers. I followed this main course with a couple handfuls of almonds, dried fruit, and M&M's. I desperately wanted to build a fire, change into dry clothes, and wait for the sun to break out, but I was now committed to meeting Walt Grayson at the Shubuta ramp, and I had no idea how many snags and obstructions I would find on the way there. I had to push on for at least most of the afternoon.

Ernest and Travis were more comfortable, I was sure. If they were paddling at all, they were dressed in rubber boots, long pants, wool shirts, and parkas, staying warm and dry while they moseyed down the river with hardly a care in the world. Ernest would be appalled at the thought of wearing wet shorts and neoprene booties in such conditions, and has already made such remarks in part 1 of this book. While it is true that my attire might not be ideal for such a cold, windy day on a woodland river, the situation would be reversed if we were paddling open water. Out there, wind like this would produce breaking waves that would quickly soak the occupants of an open canoe, while I would be warm and dry, sealed in my watertight cockpit. Admittedly, squeezing into the cramped, wet kayak was a pain here on this gentle river where it was unnecessary. There was always water sloshing about around the seat, picked up by my feet each time I got in or out and dripped in by the paddle blades if I paddled without the spray skirt sealed.

Speaking of clothing, Ernest and I have distinctly different philosophies when it comes to river attire. His approach is that of the early European explorers and colonists who regarded nature as something to shield oneself against and conquer and overcome. I like to go native, wearing the minimum conditions require and sometimes nothing at all in such places as some of the uninhabited islands I have camped on. The differences in our clothing choices are apparent from the ground up: Ernest always wears heavy boots—usually waterproof rubber ones—while canoeing, even in the heat of the summertime. He has a strong aversion to getting his feet dirty, muddy, or even wet. I, on the other hand, go barefoot as much as possible, except in the coldest conditions. In the kayak I wear the neoprene river booties not to keep my feet dry, but to provide a cushion for my heels as they rest on the inside of the hull. I love the feel of riverbank sand under my feet or mud squishing up between my toes. Isn't getting back to nature what we're out here for anyway?

Ernest also has some inexplicable, deep-seated prejudice against men wearing shorts. You'll never see him out on the street in anything but long pants, and even on the river it is the rare occasion when he goes bare-legged in a pair of swim trunks. I hate long pants and wouldn't even own any except for the brief part of the Mississippi winter when it's too cold for shorts. When the thermometer is reading 50 degrees or above, I'm in shorts, whether I'm working, traveling, or just hanging around at home. Long pants are too restrictive of movement, too hot, and too bulky to pack. I usually take one pair on a camping trip just for times when the mosquitoes or other insects are fierce and I need them for protection. The best long pants I've found are the standard-issue military BDUs (battle dress utilities). These pants are lightweight, durable, quick drying, have lots of pockets, and feature drawstring cuffs to keep crawling things from finding their way up your legs.

Our differences are not as apparent when it comes to shirts, except that Ernest more often than not favors long sleeves, for protection from the sun. Having spent so much time on open water in both kayaks and sailboats, and being a beach lover at heart, I'm more accustomed to getting a deep tan than trying to protect myself against the sun. But having survived all my trips up to this point and, I hope, grown a bit wiser, I'm more likely to apply liberal amounts of sunscreen than I used to and to cover up in the hottest part of the day.

I don't much care for hats, but you'll never see Ernest without one. Even in town he has his basic baseball cap, and on the river some sort of wide-brimmed sun hat. Just before this trip he ordered yet another expensive hat from one of his mail-order catalogs and modeled it for me before we left, asking my opinion. I mostly kept this to myself, but tried to convince him to save his money and pick up a hat like mine for $7 at the army surplus store: the standard camouflaged "boonie" hat that can be rolled up and stuck in a pocket when not in use. I've found that it offers practical sun protection, and the floppy brim does a good job of shedding rain that would otherwise get in your eyes, but it is by no means a waterproof hat. For rain I had my lightweight backpacker's parka, with a hood that could be pulled down over the flexible boonie hat.

Ernest mentioned how he approved of the way Travis dressed, since they were both similarly attired in their khaki pants, long-sleeved plaid shirts, and boots. He hinted of his disdain for those of us in our shorts, neoprene, and high-tech boats as if we were somehow less the genuine article. But try as they like to get this "salt of the earth look," they could never blend in with the locals who live along and fish in these rivers. The canoes, no matter how beat-up or cheap, are a dead giveaway. Few self-respecting good ole boys are going to fish out of a canoe, much less paddle one on a long trip. And a kayak? I might has well have landed in a spaceship, so why even bother trying to fit in?

It didn't matter what I looked like anyway. There wasn't a soul on the river to see me go by or witness my antics as I clambered back into the wet kayak after lunch and hurriedly paddled on in an attempt to generate some body heat. As long as I could stay sealed in the boat, paddling steadily, I could get warm, but with the inevitable snags that I knew were ahead, I would have to keep stopping and opening the spray skirt for photography. The amount of distance I was able to cover rapidly in between stops made up for the discomfort of a wet kayak seat, however, and the GPS indicated average speeds of 5½ to 7 mph when I was paddling. Ha! Try that in a canoe! I could make 40–50 miles per day even on a relatively sluggish river like the Chickasawhay if all I had to do was just paddle and look at scenery like Ernest and Travis were doing.

I spent two more hours after my lunch break cataloging more obstructions and enduring relentless cold wind. The highlight of this stretch was a sandbar on the west bank overhung by thickets of large saw palmettos. These tropical-looking plants warmed my heart despite the bitter cold, and I stopped to investigate and take pictures. Saw palmettos are much more common near the coast, and especially in the lower reaches of rivers like the Pascagoula, where they grow in a dense understory in bottomland hardwood swamps. It seemed odd to find this isolated thicket of palmettos so far removed from the coastal environment, but I've seen them here and there along the Pearl and other rivers farther inland.

The saw palmetto was an important plant for both Native Americans and early settlers. The palm frond leaves were used just as other palms in the tropics for making waterproof thatched roofs, and the leaves also yield strong fibers that can be twisted into cordage and rope. The saw palmetto is also a dependable food source, yielding starchy blue berries that can be eaten as is or added to soups or dried to make meal. And year round, the inner heart found by cutting into the short trunk at the base of the fronds is edible raw or cooked. These "hearts of palm" found in the saw palmetto are not quite the delicacy produced by other palm species, such as the cabbage palms of Florida or the coconut palms of the tropics, but they were surely welcome additions in a time when hunting and gathering was the only option. Using a machete, I have cut out saw palmetto hearts on other trips and found the edibility varies from region to region. On our previous Chickasawhay trip in 1993, I gathered some after Ernest and I made camp one night and sautéed them in the skillet to go with our main course. Ernest was not impressed, that was obvious, as these particular palm hearts were decidedly bitter, so I have not offered them to him again. The palmettos I've found along Black Creek were much better for some reason, but Ernest probably wouldn't believe it. He's always been doubtful of my experiments in living off the land, but the fact is obvious that people did it for millennia before we started getting everything at Wal-Mart.

Ever since I started roaming the wilds, I've always been interested in learning about what natives and early explorers ate when they had to survive off the land. I almost always carry a handy little field guide called *Edible Wild Plants of North America*, and with it as a reference I have sampled scores of plant foods from Canada to south Florida and from Appalachia to the Rockies.

I had no desire to try palm hearts again today, however, as I had plenty of food and just wanted to get to camp where I could warm up. But seeing palm leaves in this cold forest reminded me that the sunny coast was somewhere downriver, and made me want to paddle that much harder.

Back on the water, I came upon another rapid that was longer and trickier than the first one below De Soto. This drop featured a long downhill run of about 200 feet with 2-foot standing waves in the strongest part of the flow. Once again I was able to avoid getting soaked by skirting along the fringes of the main chute. At a lower water level, this would not be possible.

Downstream of this rapid, I began to look for a place to camp, knowing I was within an easy half day's paddle to Shubuta. It was time to dry out, warm up, and take it easy for the first time since the trip began. I found the perfect spot by 3:00 p.m., and felt I could now afford to indulge myself in an early camp for at least one night. Warmer weather and sunshine were surely ahead, so there was no point in being out any longer on such a cold, gray day.

This campsite for my fourth night out on the river was spectacular. A narrow branch just wide enough to paddle the kayak into entered the river from the east bank, at the foot of a steep bluff that afforded a commanding view of the river from a height of 50 or 60 feet. A sandy deer trail led to the top of the bluff, where the terrain leveled out in an expanse of mature hardwood forest fringed by a canebrake on the edge of the bluff. A giant cypress tree draped with Spanish moss stood at the mouth of the branch, once again adding that unique hint of southern swamp to the hardwood highlands that characterize this part of the Chickasawhay. It was a job to haul all my gear to the top of that bluff, but the view was worth it. Grasses and forest litter covered the ground, so I could pitch my wet tent without the added hassle of sand sticking to everything. I slung a spare rope between two trees and began hanging out my wet clothes to dry. The cold wind alone would have done the job in short order, but much to my delight, the cloud cover began breaking away and sunlight filtered down through the canopy of trees.

Since I had plenty of time, I built a fire so that I could warm up and brew some hot tea. This was the sort of place I could make a base camp and spend days exploring the surrounding woods, as I used to frequently do on my excursions to Black Creek, another tributary of the Pascagoula. I figured it wouldn't hurt to practice my old woodcraft skills, so I left the stove in the kayak to conserve my limited supply of propane for the evenings when I would be traveling long hours and not want to be bothered with building a fire. It's really quite simple to cook on an open fire without any special equipment if you know a trick like the one I was taught by three backcountry natives on a remote jungle beach in the Dominican Republic.

I was circling the coastline of the huge but remote Bay of Samaná as part of my much longer kayak journey in the Caribbean when I met these masters of fire building as they stumbled into my campsite one morning. They were carrying string bags of taro roots on their shoulders, traveling overland on foot to take their produce to market in the town of Samaná. Striking up an awkward conversation in Spanish, I tried to explain what I was doing there in a kayak, and as a gesture of hospitality they offered to share some of their vegetables if I would only loan them a cooking pot to boil the water. I had been cooking on fires for months myself, having long since worn out two expensive camping stoves and thrown them into the ocean in disgust, but my method was crude compared to the eloquent simplicity of these woodsmen. They had water boiling as fast as any modern would-be outdoorsman could unpack and set up a high-tech camping stove.

The method is ridiculously simple. Use a machete to cut three small pieces of green wood about an inch and a half in diameter and one foot long. Sharpen one end of each of these and drive one into the ground at about a 60-degree angle in the spot where you plan to build your cooking fire. Hold the cooking pot in place to gauge where to drive the other two stakes, then drive them to the depth of the first one, so that all three are the same height, the tops of them wide enough to support the pot at three equally spaced points, forming, in effect, a sturdy tripod that will hold even large, heavy pots filled with water and food. Cooking is done with nothing more than a pile of dry twigs about the diameter of a pencil. Stack these up under the tripod and ignite with leaves or other tinder and keep it fed with a steady supply of these twigs and you'll have water boiling in no time.

The possession of a metal pot in the wilds is a true luxury that we take for granted. A pot enables one to utilize all sorts of wild plant and animal foods that would be inedible without boiling. Primitive people in the past and in parts of the world today use all sorts of ingenious ways to cook without utensils, but as one well-known survival expert and author wrote, if given the choice of just one item from civilization to take on an extended stay in the wilderness, he would choose a metal cooking pot over all else, even a knife.

The vegetables the Dominicano bushmen cooked for me that day on the Bay of Samaná required a full hour of boiling. By this time the green support sticks of the tripod might be nearly burnt up, but you can judge the size you need by how much time you need to cook. On my bluff campsite overlooking the Chickasawhay River, one such setup lasted for afternoon tea, rice and tuna dinner, after-dinner hot chocolate (spiked with Captain Morgan), and morning coffee, followed by oatmeal. All this cooking required little wood, just the snapped-off twigs from the ends of larger branches that I used to build a separate, second fire for warmth and light.

Fire was a welcome companion in my camp that night after spending the entire day feeling miserably cold. I knew things were only going to get better from here on downriver; the promise of sunshine that afternoon assured me of that. It felt good to be here, high above the river in the dark embrace of the night forest as the fire crackled and popped and the river murmured softly far below. I had a smug feeling of being at home out here and confident of being able to make myself comfortable in any wild place, unlike so many folks in our modern, high-tech world of disconnection from nature. But I knew that our so-called modern "adventures" were nothing in comparison to the real explorers who blazed trails through the wilds of this continent long before the maps were made, rivers named, and roads built. It seemed laughable that we were actually getting media attention for a little two-week jaunt down a relatively tame river in rural Mississippi. Some people hearing of our trip were reacting as if we were off to the ends of the earth, so far removed are they in their comfortable lives from anything remotely resembling hardships or discomforts.

We were paddling 200 and something miles for fun, and downstream at that. In the big scheme of things it was not that long ago when people *had* to paddle down rivers, *and* back up them if they wanted to travel at all in many places. The rivers were the highways through terrain that often afforded no other practical means of travel, and everyone from the pre-Columbian natives who lived in this region to the French explorers and later settlers routinely made journeys up and down the rivers that would seem epic by today's standards. Everyone knows about the famous frontier travelers like Lewis and Clark and Daniel Boone, but countless others who never made it into the history books lived lives of adventure that would be hard to fathom today. The first European explorers to see the upper reaches of the Pascagoula and these tributaries we were paddling were the French who arrived with d'Iberville in Biloxi in 1699. They built their own boats in primitive conditions, using hand tools, and then set out on tentative expeditions to investigate all the big river systems that entered the Gulf of Mexico in that region, including the Mississippi River and the Pascagoula. These Frenchmen paddled the rivers upstream in boats that were surely heavier and less efficient than our modern canoes and kayaks. They and others like them penetrated deep into the interior wilderness of North America not knowing what they would find and having no idea how far the rivers ran or what hostile native tribes they would encounter along the way.

People today, if they venture into the wilds at all, are never far from the safety net of civilization, and most are connected to the umbilical cord of society with modern communications devices such as the cell phones we were carrying. Still, to do what we were doing is relatively uncommon, even if it was not some great adventure. Amazingly, most people these days drive over bridges across rivers like the Chickasawhay and seldom even glance down, much less wonder or care where the river goes.

I woke at daybreak in my camp atop the bluff to the sound of birdsong enveloping me from the surrounding trees. I rekindled the fire so that I could spend the cold part of the morning lounging in camp. I had the great luxury of time this morning, and I planned to enjoy it to the fullest. I made myself comfortable in my folding camp chair on the ground near the fire, and sipped several cups of coffee while writing in my journal and watching the rays of the rising sun slowly begin to filter into the forest through the smoky mists that hung over the river. I was content to know that the return of the sun meant a return of more seasonal temperatures.

I cooked a batch of pancakes and broke out the maple syrup for breakfast, then decided to do a bit of exploring before packing up and leaving. The hardwood flat beyond my campsite seemed to go on and on, and I spent a hour or so hiking, marveling at the size of the old-growth timber that someone had forgotten to cut, at least so far, and staying ever mindful of my route, as it would be all too easy to get turned around in such a trackless forest. It would have been nice to have the whole day there and to return to the camp that night in order to really feel a connection with the place, but this trip was not to be like that. Reluctantly, I began to carry my gear down the steep bank, and an hour before noon I was ready to paddle out of the still-water slough that had hidden my kayak for the night and reenter the river to make my appointment at Shubuta.

Lagoons and quiet oxbow lakes are found along the lower Chickasawhay as the terrain becomes swampier where the river nears its confluence with the Leaf.

13

With three hours available to make a mere three miles to Shubuta, I didn't have to worry too much about paddling. My first priority was to find a sunny sandbar somewhere along the way where I could stop and take a bath in the river—never mind that the water and air were still too cold for comfort. I wasn't about to go in front of Walt Grayson's TV camera until I had washed my hair and changed into a clean T-shirt and shorts. This was the fifth day of the trip and I had already taken one river bath two days ago, before the bitter cold front rolled in. Unlike some river travelers paddling a sister river who therefore don't necessarily have to be mentioned by name, I like to take baths often when in the wilds. I take one every day, in fact, if it is at all feasible, just as I would at home. There's no excuse not to, really, when one is traveling all day on such a clean source of bath water and a quick bracing dip is so convenient and rejuvenating.

I found the perfect stopping place on an east bank sandbar that was exposed to full sun, but seemed isolated enough to ensure privacy, as if that were even an issue on a river that I had traveled so many days on without encountering another user. Since I had a bit of time to kill, I first got out my small solar panel and propped it up so it was facing directly into the sun and connected it to my cell phone. I had used such a solar panel with great success on my Caribbean kayak journey, but the difference was that on that trip I often stayed put on one beach for two or more days at a time, so the solar charger had time to do its work. Back then it was a marine VHF radio I was charging, not a cell phone, but I hoped to at least keep my phone battery hot enough to make an emergency call, even if I didn't have the luxury of being able to call Michelle again.

This done, I stripped and waded into the cold river. It immediately took my breath away when I went in over waist deep, so there was nothing to do but get it over with quickly. I dunked my head and then washed my hair and body with biodegradable shampoo and submerged again to rinse. It's just that simple to bathe in a river, and so refreshing it's worth any amount of pain from the initial shock of cold water. I shook the excess water out of my hair and walked around on the sandbar to dry in the sun. There was time for a leisurely lunch, so I dug out all the possible ingredients for a meal that required no cooking and spent a good hour on the sandbar to give the phone some charging time.

Just downstream from the sandbar, the river took a sharp bend back to the west, and the sand on my side gave way to a sheer bluff that towered maybe 60 feet above the water. At its base was a gentler slope that was littered with boulders, some almost as big as cars. Such rocks are unusual in Mississippi and beckoned to be investigated. I put on some shoes and got out my SLR camera to take some pictures, then picked my way along the steep bank to the rock field. The boulders appeared to be made up of mostly limestone. Closer inspection revealed that they were full of marine fossils, from small clam shells to perfectly preserved sand dollars. The sand dollars were the most prevalent, sticking out like blades from the broken edges of the rock. I adjusted my zoom lens and camera settings for some close-up photography, and just as I was about to begin taking pictures, the low battery indicator on my Nikon began blinking. I couldn't believe it. I had brought dozens of spare AA batteries for the Kodak digital camera and the GPS and Mini Maglite, but the one device I did not bring a spare battery for was this new SLR camera, which I had just purchased recently and which came with a long-life lithium battery. I had not wanted to spend $30 on a spare battery I didn't think I would need, and now I was going to pay for it. I still had almost 10 rolls of color slide film that were now useless. High-quality photos were essential to illustrate our book and especially any spin-off magazine articles Ernest and I might decide to do. I had even bought the extra zoom lens I already mentioned for the chance wildlife photograph. Now I could kick myself. All I had was the cheap digital camera, and though I had bought two extra memory cards for it, I could not afford to use disk space photographing scenery, as I still did not know how many river obstructions lay ahead and had to save room for shooting all that I found if I expected to get paid by the Pat Harrison Waterway District. Where would I possibly get such a specialized battery out here in the boondocks of southeast Mississippi? I didn't know what kind of stores might be in Shubuta, but it was possible a drugstore might have the right battery. I put away the camera with disappointment and picked through the rocks looking for samples of the sand dollars I could take with me.

It might seem preposterous to find sand dollars over a hundred miles inland as the crow flies, from the Gulf of Mexico, but fossils of marine invertebrates such as starfish, mollusks, and corals are common in the gravels and rock outcrops across most of Mississippi and neighboring states. The entire area was covered by a shallow sea from 320 to 480 million years ago, and more impressive evidence of this is occasionally found, such as huge teeth from long-extinct giant sharks and skeletons of large marine mammals.

Sand dollars are from the class of marine animals known as echinoids—spiny-skinned creatures. They are closely related to sea lilies, sea cucumbers, starfish, and sea urchins. When alive, the fragile disk that is the skeleton of the animal has a velvety covering of minute spines and is darker in color than the often sun-bleached sand dollars found on the beach. They live on top of or just beneath the surface of sandy or muddy areas, burrowing or slowly creeping through the sand using the spines on the underside. Hair-like cilia cover the spines and move food in the form of plankton and organic particles to the mouth opening, which is in the center of the star-shaped grooves on the underside of the animal.

Sand dollars vary in size according to the species. The fossilized ones I found in these rocks on the Chickasawhay River averaged 2 to 3 inches in diameter, but on a scuba diving trip off the coast of Florida, I once collected several sand dollars that were 5 to 6 inches across. It's somewhat rare to find an intact one on the beach, as they are usually broken or chipped by the waves that tumble them ashore. But these sand dollars had turned to stone and many were perfectly preserved. I took off my T-shirt and wrapped a few in it for safe transport in the kayak, and reluctantly left the boulder field to head on downriver. I had about an hour remaining until 2:00 p.m.

As it turned out, I could have spent more time searching for fossils. I reached the Shubuta boat ramp early, and found it deserted. The ramp was damaged by recent floodwaters, and the river gauge marker nearest the stream had been knocked down and half buried in sand. I pulled the kayak far enough up on a sandbar at the foot of the ramp to ensure it wouldn't drift away, and then wandered around the parking area atop the steep bank. A few picnic tables and an abandoned pavilion were the only facilities.

A derelict iron bridge crossed the river just downstream from the boat ramp, and the road leading in to the parking area was now a dead end with the bridge no longer serviceable. There was a house just up the road before the entrance to the boat ramp, and I could hear the sound of a lawnmower through a line of trees that blocked my view of the yard.

Several minutes after 2:00 there was still no sign of Walt Grayson or anyone else, and I began to wonder if he had gotten my message. I went back to the kayak and got my cell phone, but when I turned it on there was no signal, even when I was standing at the highest point on the riverbank. I turned it back off and waited. By 3:00 I decided to walk up the road leading in to see if I could find someone who would let me use a phone or who could tell me if there was a drugstore or hardware store where I might find batteries for my camera. Just as I reached the road, I saw a familiar-looking Chevy Tahoe tearing around the bend in my direction, and when it slowed to a stop beside me, Walt and his wife, Jo, explained that they had gotten pretty well lost trying to find the boat ramp and had wandered around quite a bit before getting reliable directions from a local resident.

Walt Grayson is a familiar face and a household name just about everywhere in Mississippi. His insightful commentaries and distinctive voice are familiar to anyone in the state who listens to public radio or watches television. Walt has made a living out of wandering back roads, and shares his discoveries with television viewers all over the state. A fifth-generation Mississippian born and raised in Greenville, Walt got his start in broadcast journalism as an announcer for a radio station in Greenville while he was still in high school. He later worked as a weekend weather filler and travel reporter while still working full time in radio, and then became a weather anchor at WLBT-TV in Jackson in 1984. Walt has gotten away from his weatherman duties to pursue what he really loves, seeking out interesting places and people throughout the state and bringing them to the television viewing public in a way that is both entertaining and informative. His ETV television show, *Mississippi Roads*, which he both hosts and writes, showcases his advocacy for the state of Mississippi and has done much to educate state residents about the variety of interesting people and places to be found in their own backyard. "Look Around Mississippi" is a similar feature that he does for WLBT-TV, working independently, finding the stories and doing both the filming and the writing. Walt is also a writer, contributing a monthly column for *Today in Mississippi*, the magazine of the Electric Power Associations of Mississippi, and he still does radio commentaries on Mississippi Public Radio.

Walt Grayson called me about a week before we were scheduled to depart for this river trip, asking questions about the project and saying he would like to meet us at the put-in and at various points along the way. He got wind of our project thanks to Nancy Cleveland, Michelle's mother, who is a personal friend of one of Walt's coworkers at WLBT. Ernest and I had appeared before on *Mississippi Roads* back in 1999, when I was featured on a segment of the show for my boat-building hobby, which at the time had yielded a variety of small boats, from a decked canoe and a sailing catamaran to the "Mississippi backwoods drifter," a sort of double-ended johnboat I designed and built for Ernest. (This is another story I won't go into that illustrates Ernest's powerful persuasive abilities. It is the story of how he managed to talk me into designing and building for him a gorgeous, handcrafted wooden boat for a mere $700—barely enough to cover the cost of the expensive woods and epoxy. He thought it was a fair price until one of his more generous friends commissioned me to build a similar model for a much more realistic $2,500.) Although Walt was the host of *Mississippi Roads*, we didn't get to meet him that time. A film crew from ETV came to Prentiss to interview me and ask questions about the boats, and we took Ernest's "drifter" and a fishing boat I built for my dad out to a local lake to shoot some in-the-water footage.

Walt told me he planned to use our Pascagoula story for "Look Around Mississippi" rather than *Mississippi Roads*, which was already filmed in advance for several months. We made arrangements to meet in Taylorsville on the Saturday of our departure, and as Ernest has already mentioned, Walt filmed us as we launched Travis's and Ernest's canoes and then interviewed Ernest and me under the bridge. Camera-shy, Travis declined the opportunity to be interviewed. Walt and his wife, Jo, had to get back to Jackson early that afternoon, so they didn't have time to follow Eddie McCalip and me to my launch site on the Chickasawhay at Enterprise. This meeting at Shubuta would be Walt's first look at my kayak in the water.

Walt's wife, Jo, was just as enthusiastic about our trip as he was. She told me that she goes with him every day on his quest for stories. They both seemed to love doing this, and why not? Day after day of getting paid to cruise back roads and talk to interesting people—who wouldn't envy a job like that? Walt was always on the lookout for a story, and getting lost trying to find the Shubuta boat ramp gave him ideas about returning to the area. He said it was one place in the state he had never been.

As I pointed out the boat ramp where I'd left my kayak pulled up on a sandbar, it occurred to me that Walt and Jo were the first people I'd spoken to since Sunday morning, when I'd had my brief encounter with the two men and the teenage boy fishing at the Stonewall boat ramp. Now it was Wednesday afternoon. It was amazing that such isolation could still be found in Mississippi and that I'd seen no one on the river despite the fact that I'd traveled almost 50 miles. But I knew it was largely because of the weather and the fact that I was out during midweek when most people were at work. I expected to start seeing a lot more people now that I was getting to more navigable water and the sun was out to stay, at least for a few days.

I told Walt and Jo about my dilemma with the camera battery and they immediately volunteered to drive back to Shubuta and see if one was available. I waited another 20 minutes until they returned, unfortunately, empty-handed. The local hardware store carried only the usual assortment of AA, C, and D-cell batteries. The storekeeper told Walt that Waynesboro was the nearest place to get the battery I needed. There was a Wal-Mart there, he said. Walt and Jo were en route to the coast after meeting me, and could not go to Waynesboro, and I would not have wanted them to. I would have to do without and it was my own fault. Walt took down the model number of the battery and said they would bring me one the following Monday, when they planned to rendezvous with all of us again at Merrill.

Walt set up his massive video camera on a tripod at the foot of the boat ramp and conducted a brief interview in which he asked me questions about what it was like to be out there alone, what I had seen, etc. He then asked me to paddle back upstream out of sight and come down the river to land as if I were arriving at Shubuta for the first time. The last scene he filmed was of me paddling off downriver, sliding beneath the dilapidated iron bridge below the ramp, and disappearing around the bend. When I later saw a tape of the program, Walt's evocative commentary that has become his trademark added a mystical appeal to the journey that made the whole thing seem like a much bigger deal than it really was.

The appointment met and the interview over, I paddled at a good pace to put some distance between myself and Shubuta in hopes of finding a relatively remote campsite. It was already late afternoon, nearly 4:00 p.m., so I didn't plan to be on the river much longer. As it turned out, I had to paddle farther than I would have liked. In the first few miles below Shubuta, the high bluffs and deep woods I'd become accustomed to were replaced by low clay banks, pine plantations, and fenced-in cattle pastures. There were sandbars in some of the bends, but all of them seemed accessible to the landowners by dirt roads or ATV paths, so I pushed on. Snags were abundant in this area, so despite the fact that I was tired and ready to make camp, I still had to stop and work at almost every bend, and this continued until the sun was getting so low I was going to have to stop somewhere or end up paddling at night.

A sprawling expanse of white sand on the east side of the river was too tempting to pass up, despite the fact that there were cattle tracks everywhere in the sycamore grove at the top of the bar. I decided to camp there. It was late, and once again, I would be leaving early in the morning. If there was a way to drive to the sandbar, it wasn't obvious, and the place felt sufficiently remote for a good night's rest.

That night, with a high-pressure system in place over the area after the passing cold front and not a speck of cloud in the sky, the universe unfolded overhead above the wide-open sandbar in a dazzling array of stars that made me feel as if I were camped in the high deserts of Arizona or New Mexico. As if in conspiracy to reinforce that idea, a pack of coyotes suddenly filled the night air with a chorus of high-pitched singing and howling, so close they had to be in the pasture just beyond the sycamore grove that backed my camp. I love to hear that sound, and the idea of wild dogs crying out in the night is enough to make any place seem remote from civilization. Like the huge populations of deer now thriving in Mississippi, coyotes have come to these parts relatively recently. They were unheard of when I was a boy growing up in Prentiss, but now are common throughout the state. They are heard much more often than they are seen, but one afternoon in broad daylight I walked right up on one in a clearing on my family's land near the Bowie River. I didn't expect to see any on this trip, as they are nocturnal for the most part, but it was nice to sit back and listen to their music, until it abruptly stopped just as suddenly as it had started. The vacuum was quickly filled by the usual night sounds: the chirping of crickets, murmuring of the river, and the occasional croak of a frog. I stayed out late on the sandbar, sipping hot chocolate, smoking a cigar, and lying on my back studying the infinity of stars within the Milky Way through my powerful binoculars.

It felt good to be here, and with each passing day I was becoming more attuned to my natural surroundings. The other world of traffic, deadlines, and work seemed quite far away. I couldn't talk to Michelle tonight, but I had used Jo Grayson's cell phone to leave a brief message on her voice mail to let her know that I was doing fine. She would worry about me on the days that she did not hear from me, but that was because she was still unaccustomed to what I was doing. With the advent of cell phones, our society has become dependent upon instant and constant communication. It is as if we cannot be safe if someone doesn't know where we are at all times.

It is true that there is a degree of inherent danger in going into the wilds solo, and as I grow older and more experienced I have learned not to take anything for granted. I am more careful about every move I make, and try not to take unnecessary chances. Even when you least expect it, things can go downhill fast when you're alone, whether in the bush or close to home. I learned this again a couple of years before in a frightening experience just a few miles from where I grew up in Prentiss.

I was riding my bike on the Longleaf Trace, the newly opened rails-to-trails bicycle trail that runs from Prentiss to Hattiesburg, when it happened. It was a hot Saturday morning in late spring, and my heart rate was up from 7 or 8 miles of steady pedaling when I felt a painful sting in the back of my neck. I reached back and swatted at it, briefly feeling the hard body of a wasp or bee as my fingers swept it away. For a few seconds I pedaled on, thinking little about it, until I started feeling weak and a little dizzy. I decided to turn around and cut my ride short. I got back up to my usual speed of 18–20 mph and one minute I was riding and then everything went blank until I woke up with my face buried in the trailside gravel. My feet were still in the toe clips and the bike was lying flat beneath me. I felt a strange, numb sensation all over my body and my throat was constricted, making it difficult to breathe. It occurred to me that I had had a severe allergic reaction to the sting, and I was temporarily paralyzed and unable to move. I don't know how many minutes passed as little by little I regained enough mobility to roll over on my back. I still could not raise my head or reach my desperately needed water bottle. From my helpless position I watched puffy clouds pass overhead above the canopy of trees shading the trail, and felt waves of pain as my breathing constricted. I began to seriously wonder if my adventures were going to end right then and there.

The poison didn't do me in, but it was a close call. I was eventually able to raise my head and reach my water bottle. A cell phone in my handlebar bag would have been mighty convenient right about then, but I had left mine at home. Finally, some other bicyclists came along and one went to call an ambulance. I recovered after a trip to the emergency room for an adrenaline shot and a series of tests.

I now carry an Epipen shot for stings, and keep an extra sharp lookout for bees and wasps. But I learned from that experience not to take anything for granted, and I realized that if technology can get you out of a tight spot like that, why not use it? So the cell phone was with me, and despite the low battery and frequent absence of a tower signal, I knew there was always a chance it just might save my life, given the right circumstances. But using it would be my prerogative out here and it was strictly for outgoing calls. I was off the hook as far as any work-related or casual chitchat calls were concerned.

Cypress trees are supported by weird aerial roots called "knees." They often form in strange shapes, such as these that look like faces.

14

Back on the river the next morning, a short distance downstream of the big sandbar, I left the piney cutover and once again entered a world of towering hardwoods, fern-and-moss-covered bluffs, and straight stretches of deep, still water. This section of the Chickasawhay turned out to be perhaps the most scenic stretch of the river. From the topo maps in the Pat Harrison report, I determined that I was about 7 miles above the Highway 84 bridge crossing at Waynesboro when I entered this fascinating part of the river.

The sound of falling water from the woods on the outside of a sharp bend drew me to the west bank to investigate. What I found was a cascading waterfall that would not have been out of place on a tropical island. The waterfall was just back from the riverbank, almost hidden in the grove of river birch and sycamore trees that grew on a narrow bench situated between the river and a tall, rocky bluff. A tributary stream plummeted over the edge of this bluff and fell a good 10 or 12 feet to a deep pool of water, before running off into the river. What made the setting so magical was the jungle-like thicket of palmettos growing at the base of the falls and hanging over the lip of the bluff. Thick mats of hanging ferns growing from the wet face of the rock added to the tropical effect. Naturally, this was just the sort of scenery I would encounter now that my Nikon camera was useless with a dead battery. I took several photos with the digital camera, as this was too good to pass up and I felt like I would have enough disk space to finish my report so long as I did not take more than 50 or so personal photos.

There was no reasonable way up to the top of the bluff without considerable risk to life and limb, so I decided to forego trying to climb it. This bluff rose even higher just downstream from the waterfall, and in places formed a wide overhang that looked like the sort of open caverns that were often used by early native people for campsites. I once found several arrowheads in just such a rock shelter near the Buffalo River in Arkansas, but though I spent several minutes searching the ground, I saw no flakes or other evidence that Stone Age hunters had camped here. If they didn't, they certainly overlooked a prime campsite. The flat shelf between the river and the foot of the bluff was well shaded by big trees, but devoid of underbrush. There was an excellent view of the river, both upstream and downstream. Once again, I had found a place where I could have hung out for a few days, but since it was only midmorning, I reluctantly climbed back in the kayak to push on.

Downstream of the waterfall, the wonders of the river didn't end. I felt as if I had been transported to a much more exotic locale than eastern Mississippi as I drifted in silence beneath more towering bluffs, all covered with moss and ferns, some with palmettos growing right out of the side of the rock. A cluster of sycamore trees with a gigantic root system drew me to the east bank. The five or six separate trunks were actually sharing a common base, and the entire root system was growing out of a vertical rock face, with huge, twisted runners that looked like those of some tenacious jungle tree clinging to the ruins of an ancient temple. I had seen similar trees in just such places at the Mayan ruins of Copán, in Honduras, but to see such a massive tree here, with a base that must have been 20 feet across, was most unexpected.

A little farther downstream, another tributary entered the river from the east bank in a series of minifalls and cascades down a boulder-strewn ravine. This ravine was spanned by a railroad bridge, not the usual steel trestle span seen most often on Mississippi streams, but rather a masterfully built masonry arch that seemed to blend right into the rocky bluffs at each end of it. I stopped for a snack on a sandbar at the entrance to the ravine and enjoyed the shade and the sound of falling water. I knew I was close to the boat ramp at Highway 84 and decided to wait and have lunch there, where I planned to walk up to the road and see how far it might be to Wal-Mart or some other store where I could find a camera battery.

When I reached the ramp, there was no one around. At the top of the hill there was an abandoned parking area and some picnic tables, and a couple of houses were in sight farther away, but the bridge crossing was obviously some distance from any stores. I decided against taking a long hike away from the river, partially because of the time it would involve, and partially because I was afraid to leave the kayak and all my expensive gear unattended. I ate lunch at the foot of the ramp and smoked a cigar while I plugged in the waypoints for the obstructions in the next 20 miles of river.

I was stuffing my notes back in the kayak and preparing to leave when a car drove down the ramp right to the river's edge and stopped beside me. The man driving the car introduced himself as Lester Wayne Walley. He was out riding around with his family and had spotted my kayak from the bridge. Lester Wayne said he was a local who had grown up in the area and knew a lot about the river. He was glad to be back home as he had just returned the day before from a 28-day hitch on his overseas job in South Korea. When I asked him how far it was to Wal-Mart, Lester Wayne immediately volunteered to drive there and pick up my battery. I gave him the money and sat back to wait. Twenty minutes later, the Walley family returned with two new batteries for my camera. Wal-Mart had a cheaper version of the same battery, so he was able to get two of those for less than the price of one lithium battery like the one the camera came with. I was ecstatic about having a functional SLR camera again, and felt lucky that I had to do without it only on the stretch from Shubuta to Waynesboro. I thanked Lester Wayne Walley and his family, and they watched as I climbed back into the cockpit of the kayak and shoved off.

Meeting the Walley family was to mark the beginning of seeing a lot more people out on the river for the rest of the trip. The next boat ramp was at Highway 63, and in the water near the ramp I passed a disgruntled-looking young man with a female companion. They were in a powerboat that looked like a 60s vintage ski boat, and the man was having major problems trying to get the cantankerous old engine to run. He didn't seem particularly happy to see me gliding effortlessly by under paddle power, so I just nodded and didn't try to strike up a conversation.

A little later that same afternoon, an elderly couple in a johnboat waved and called out to me from where they were drifting near some overhanging willows on the west bank of the river. The Chickasawhay was wider here below Highway 63 than it had been in the upper reaches, so I had to paddle closer before I could hear what they were saying.

"You ain't got a shotgun handy, have you?" the man asked, as I approached.

"No. Why?"

"There's a cottonmouth moccasin over there next to one of my lines that's as big around as your leg." The man pointed in the direction of some dense willows. I could see a trotline dangling into the water, but it was too thick to make out a snake.

"I've got a .45," I said, not that I wanted to shoot any snakes with it—I just had to let this backwoods couple that looked like they had fished out here all their lives know that I had a gun. My kayak was strange enough, but having a pistol along would change their perception of me from an urbanite outsider to a passable excuse for an outdoorsman. I didn't carry the .45 for protection against snakes or any other wildlife, of course. Guns are useless against snakes, because the ones that bite you are the ones you don't see until it's too late. I carried the gun mainly for peace of mind and in case of the unlikely event that I would have a run-in with a dangerous animal of the two-legged variety. I've never had trouble on the rivers of Mississippi, but you never know. At a remote bridge on the Leaf that Ernest and Travis had probably already passed by now, there was a much publicized double murder a few years before. As usual on my river trips, the .45 had yet to be taken out of the waterproof bag I carried it in, and I wasn't about to pull it out to shoot at a snake.

"You probably couldn't hit him with a pistol. You'd just run him off."

"We're gonna go back and get a shotgun from our camp," the woman said.

"Yeah, we gotta make sure we got him," the man said. "We'd a run right up under him if we hadn't been paying attention."

"Where is he?" I asked. "I want to get a look at him."

They both pointed at the branch where they said the snake was and followed as I slowly crept forward in the kayak. I got out my Nikon and changed the lens to the 300 millimeter zoom. There was a limit as to just how close I planned to get to a big cottonmouth. The snake was amazingly hard to see, even with two people who knew where it was pointing it out. Finally I spotted it from a distance of about 10 yards, and through the telephoto lens I could tell that it was indeed a venomous cottonmouth, and looked to be about 5 to 6 feet long. Cottonmouths are exceptionally thick-bodied snakes for a given length, and as the fisherman had said, this one looked as big around as a man's leg. Its coils formed a massive brown knot, draped over a forked willow branch and blending in with roots, driftwood, and other debris tangled in the thicket. The snake was right over the suspended trotline, and a less observant fisherman could have easily passed right under it without even seeing it.

Mississippi has its share of venomous snakes, and is home to all four of the poisonous snakes found in North America: the rattlesnake (of which there are several different species), the copperhead, the coral snake, and the cottonmouth. The cottonmouth is a water-loving pit viper that usually averages 3 to 4 feet long, but 6-footers, while rare, are not unheard of. The cottonmouth gets its name from the distinctive white interior of its mouth, which it displays along with bared fangs when threatened or while taking up an aggressive and threatening posture of its own if it feels its territorial rights have been infringed upon. It's not uncommon for a cottonmouth to actually move toward a human intruder, as if to attack, but most of this is just bluff and avoiding a bite is a simple matter of having the sense to leave snakes alone. Even the fearless Australian snake catcher Steve Irwin, known to most as "the Crocodile Hunter," has a lot of respect for the cottonmouths of North America, admitting that they are difficult snakes to handle. I used to catch a lot of nonpoisonous snakes myself, but I never had the least bit of temptation to mess with a cottonmouth. I would have liked to see the open mouth of this monster specimen, but I didn't want to get close enough to prod it with the paddle, especially since it was coiled in an advantageous position well above the level of my kayak seat.

As we were talking about the snake and I took what pictures I could get, another fisherman in a johnboat motored up and showed us a huge alligator snapping turtle he had caught on one of his catfish sets. The locals around here call snapping turtles "loggerheads," but that name is correct only for certain species of marine turtles. Luckily for the snake, still coiled up unperturbed on the willow branch, this fisherman didn't have a shotgun with him either. As I paddled away, I sincerely hoped the big cottonmouth would leave the branch and disappear before these fishermen returned to kill it. I agree with Ernest, and unlike Travis, I will kill a venomous snake if I find one near human habitation and feel it is a threat to someone, but out here in the wilds along the river, I believe all snakes should be left alone.

A couple of bends farther downriver I was once again paddling in silence and caught a wild turkey by surprise as it walked along the edge of the river on a big sandbar. It was a bearded gobbler, and while I fumbled to get my camera out of its waterproof bag and ready for a shot, the turkey nonchalantly strutted up the steep sand drift and disappeared into the woods. Getting good wildlife photos requires a whole different mindset. To succeed, I would have to be out here for no other reason. As it was, all I could hope for was the occasional lucky shot.

Another missed opportunity the same day was the 6- or 7-foot alligator I paddled within 20 feet of before seeing. The alligator was sunning on a pile of debris in a bend of the river, and just as I realized what I was looking at, it took to the water with a deafening splash. I knew I was more likely to start seeing alligators the farther downstream I paddled, as the river becomes bigger and the habitat swampier and more suited to the big reptiles.

Like the whitetail deer and coyotes, alligators were rare in the wild in Mississippi when I was growing up. Aided by the Mississippi Department of Wildlife, Fisheries and Parks, alligators have made a remarkable comeback in the state and are now common in practically all the rivers, lakes, and coastal waters of south Mississippi. Large populations of alligators are found throughout the southern states, in numbers that are causing problems in Florida and that have led to legal harvesting in Louisiana. Mississippi has yet to have an open season on alligators, but this will likely come as the population of the reptiles continues to grow. Run-ins with humans are more common, and as more alligators survive to reach adulthood and grow to large sizes, attacks on humans are a possibility. Though not considered nearly as dangerous as crocodiles, American alligators have been known to kill and to eat humans, and several deaths from alligator attacks have occurred in Florida in recent years. Alligators certainly have the equipment in the form of teeth and powerful jaws to take on most any animal that enters their watery habitat, to which they are supremely adapted as master predators. I don't worry about them when kayaking or canoeing, but after a few close-up sightings in recent years I have decided to refrain from swimming in some of the exceptionally swampy places I used to spend a lot of time splashing around in.

The lower reaches of the Chickasawhay were quite different from the river I had spent the first 5 days on. Tall bluffs were replaced by broad sandbars, and pine trees were much more common in the forests here than in the hillier country upriver. Cypress and willow were abundant along the banks, and the river itself spread out wide and shallow, the water a muddier brown than in the deep pools of the upper reaches. Sandbars in practically every bend assured that I would have no trouble finding places to camp, but solitude was less likely in the area as meetings with fishermen using the river became frequent.

Billy Mashburn, a commercial fisherman who stopped to chat on the way back from running his lines, showed me two 10-to-12-pound striped bass he had caught that morning. We talked about the snags in the river and he pointed to the folded-in bow of his 16-foot aluminum johnboat, telling me that he had hit a stump in the river while running at night, breaking three ribs and nearly getting thrown out of the boat. He was, of course, in favor of the de-snagging of the river, at least the main channel, and seemed happy to learn that I was gathering data for the Pat Harrison Waterway District. Many of the fishermen I talked to knew about the proposed project, though most doubted it would actually ever be done.

On Friday night I camped about 2 miles upstream of Old Avery Landing, leaving 2 days to cover the remaining distance of about 52 miles to Merrill. I could relax a bit now, knowing for certain that I could make Merrill by Monday morning with little effort.

The next stretch of river upstream of Leakesville seemed delightfully wild, despite the fact that I was traveling it on a beautiful Saturday morning when it should have been buzzing with fishing boats. Instead, the only sound was the whispering of wind in the tall pines and the ever-present music of myriad birdcalls emanating from the forest. Swallow-tailed kites wheeled overhead, often close enough for photos, so I stopped frequently to get out the camera. There were also large flocks of vultures circling the river in this stretch, and at one point I paddled close to a group of them tearing at a dead alligator snapping turtle, which was upside down on a sandbar. As I paddled, the vultures eyed me as if wondering if I might be their next meal.

"Not today, buddy," I told one that circled especially close and swooped within 15 feet overhead.

I stopped often in this seemingly pristine stretch of river. While to some it might have been less scenic than the rocky upper reaches, to me it had an appeal that was equally fascinating. The forest here was mature and primeval feeling. The sandbars were impossibly white, and devoid of signs of humans. The only tracks were of beaver, raccoon, mice, and heron and other birds. I sat on some of these sandbars for as long as an hour, drinking in the silence and imagining that this was much the way the river had always been.

As I pondered the illusion of being the only human on the river, I thought of how confusing our definitions of wild places can be. The term "wilderness" is frequently tossed around to describe big areas of uninhabited lands, but these river-bottom woodlands, as big as they seem while one is floating down the river, cannot accurately be described as "wilderness."

Most of the remaining wilderness areas in the lower 48 states are out west, in the lonely reaches of mountains and deserts that stretch from Canada to Mexico, and in many places are quite pristine and untouched. In his book *The Big Outside*, radical environmentalist Dave Foreman catalogs state by state the untouched areas left in the lower 48 that can be called wilderness. By his definition, wilderness means "roadless," and his criteria, especially for the West, is that the roadless area be at least 100,000 acres in size. Clearly few areas east of the Mississippi meet these criteria. Notable exceptions are the Everglades National Park, the Okefenokee Swamp, parts of the Appalachians such as the Great Smoky Mountains National Park and the Nantahala National Forest, and parts of the Adirondacks in upstate New York. There are other areas in the South that could almost make the cut. The Atchafalaya River basin in Louisiana is one, at over 800,000 acres, but oil field canals and commercial barge traffic in the basin make it quite different from wilderness areas such as those found in the West.

Over the years I've spent quite a bit of time in America's wilderness areas, including numerous solo backpacking trips in New Mexico's Gila Wilderness and Montana's Selway-Bitteroot Wilderness. These are places where people are truly visitors and signs of human passage are minimal if they exist at all. In the great silence of the remote valleys hidden away in such places, a lone traveler can truly feel as if he is the only person on earth.

Here in south Mississippi, that feeling is hard to attain for long, as there is always a road, a clear-cut, a power line, or a house to be found if one sets out walking in any direction for even a mile or two. Though there might be many hundreds of thousands of wooded acres along these rivers, it can hardly be called "wilderness." Some call it simply "the woods," "the boondocks," or "the swamp." I think a better name that pretty well sums it up is "the bush." This is a common term almost everywhere in the English-speaking world but North America, with the exception of Alaska and Canada, where it is often used. In places like Africa, Australia, New Guinea, and New Zealand, any place out beyond the habitations of civilized humans is simply referred to as "the bush." I think Mississippi's backwoods and semiwild river bottoms could also be summed up more accurately as simply "the bush." This broad term eliminates the need for petty distinctions such as how many acres are "wild," lack or presence of roads, and ownership status of undeveloped lands. Certainly all the country I passed through on my descent of the

Chickasawhay River could be called "bush," despite the fact that it all consisted of privately owned tracts of land and no federal or state designated "wilderness areas."

The idea of being out in "the bush" alone for several days was an appealing notion to me. Okay, so maybe the Chickasawhay and Pascagoula rivers do not pass through the heart of a "howling wilderness"; there are signs of humans, and other people are encountered on the river, but in south Mississippi there is certainly enough wild land to do like Crocodile Dundee and "go bush" for a while. A person could disappear in these big woods if he or she had a mind to; in such expanses of river bottom swamp it would be awfully hard to find someone who did not wish to be found.

The feeling of remoteness came to an end later on Saturday afternoon as I drew nearer to the town of Leakesville and its busy boat ramp. The weather was fine and weekend fishermen were out in force. I passed several boats in the last few miles above the ramp, though no one was particularly friendly and I was mostly met with hardly a nod and bewildered stares as I slipped past in my skinny boat. I stopped briefly at the boat ramp and talked to a local man who was pulling his boat out of the water. He had just received a citation from the game warden for fishing without a license. He told me the warden was downriver, so as I paddled on, I expected to meet him sometime before I made camp. I was a little concerned about this, as several people on the river had told me that it was now illegal to camp on sandbars without the landowner's permission. Would the game warden make me pack up and leave if he saw me camping? I wasn't sure.

Below Leakesville, I was disappointed to see an endless succession of riverside camps. There seemed to be one overlooking every good sandbar, and most were occupied with families out to spend the weekend on the river. I paddled on until nearly dark and at last found a bar that was out of sight of any camps. I still had not passed the game warden, so I pulled the kayak up and unloaded my stuff. Just as I was erecting the tent, a boat approached from downriver and I saw the telltale blue law enforcement light mounted on the bow. The officer slowed down and then motored straight over to run the bow of his boat up on the sand next to my kayak.

A friendly greeting and a big smile allayed my fears that I had run afoul of the law by camping there.

"Are you one of those fellows I read about in the paper? The ones that are paddling to the coast?"

"I guess so," I said. I didn't know about any newspaper coverage that would have reached this neck of the woods. "What paper?" I asked.

"The *Mississippi Press*. There was a big article in it last Sunday."

Of course, that made sense. A reporter from Jackson County's *Mississippi Press* had interviewed Ernest and me over the phone before we left. I just didn't realize the circulation was so big as to reach this far inland.

The warden, who introduced himself as Doug, assured me there was no problem with me camping on sandbars.

"The problems are caused by big groups of drunks that come out here and party. They trash the sandbars and have no respect for the landowners. I got a call recently from a landowner who asked a group like that to leave, and they cussed him out and ran him off his own land. That's when I get involved. But most people don't mind folks like you camping for the night. If somebody asks you to leave, just pack up and paddle around the bend to the next sandbar. All you have to do is show some respect to the landowner, and nobody will have a problem with it."

Sunday was my last full day on the Chickasawhay River. The stretch below Leakesville continued to disappoint me, as there were more camps, people out riding sandbars on four-wheelers, and numerous fishing boats speeding up and down the river. Doug, the game warden, had told me I would see a dead alligator someone had indiscriminately shot, and sure enough, I found it: a 7- or 8-footer hung up on a snag in the middle of the river.

The sense of remoteness returned only after I passed under the twin spans of the Highway 98 bridge. In these final few miles of its run, the Chickasawhay was once again a world of big woods and pristine white sandbars. I stopped at 4:00 p.m. on a particularly beautiful bar, less than a mile from the confluence of the Leaf and the beginning of the Pascagoula River. I would make camp here, and call Ernest at 6:00 p.m., our prearranged time, and see where he and Travis might be. I suspected they were also in the final few miles of the Leaf. But when I called, I was surprised to learn that they were already in Merrill, and had been there most of the afternoon. I should have known, considering how Ernest likes to push while on a river trip. Ernest tried to persuade me to pack up my camp and paddle on down to join them, saying there was a good place to camp and access to a hot shower in the Nature Conservancy office. I declined. I had my heart set on one more night of solitude, and this campsite was as good as they get. There would be the entire following week to spend in the company of Ernest and Travis, so I stayed put until daybreak.

Travis, left, and Scott study map before setting out after lunch on a sandbar.

PART THREE

UPPER PASCAGOULA RIVER

ERNEST HERNDON

Scott Williams explores cypress swamp in Charles M. Deaton Preserve near Merrill.

15

Black-bearded, wearing a black shirt, shorts, and river sandals, Scott looked rugged and swarthy after more than a week paddling the Chickasawhay River solo. We talked excitedly about our experiences—of snakes, gators, fishermen, mosquitoes, rain, cold, wind, and current—in other words, condensed versions of the first two parts of this book. Travis showed up and joined in. Scott claimed his ultrasleek wooden sea kayak had performed admirably, though of course I knew all too well how miserable it must have been as he hunkered down in the cramped cockpit, struggling to record every snag on the Chickasawhay River—all while Travis and I moseyed down the Leaf, pausing for naps and espresso.

We walked up to the Nature Conservancy office, where manager Becky Stowe had arrived for work. She looked hale and outdoorsy with blond hair, olive shirt, jeans, and sandals. Soon her husband, Read, a white-bearded anthropologist in khaki shirt, jeans, and boots, arrived in a Jeep, accompanied by a half-Jack Russell terrier, half-feist named Russell.

Read was a retired anthropology professor who now ran a consultancy business called Archaeological Services, and he's conducting surveys on the two Nature Conservancy preserves here. "We think there were a lot of sites here," he said. "There are sites that date from the late Paleo-Indian period, 7500 B.C., up to the historic period."

Read's specialty is dugout canoes, and he's seen good specimens taken out of the Pascagoula River and tributaries. Indians paddled downstream and poled back up. A typical dugout was 20 feet long, carved from a single cypress or cedar log, and quite heavy. "You don't portage these things," he said. "When you're done with it you sink it . . . If you've got a portage and you've got to go overland, you don't want someone to take your canoe." Sinking a log canoe also kept it from drying out and cracking.

In an article in the December 1974 issue of *Journal of Alabama Archaeology* entitled "A Preliminary Report on Four Dugout Canoes from the Gulf Coast,"

Read went into more detail: "During late prehistoric and early historic times the rivers, creeks, and bayous of the Gulf Coast served as highways for the aboriginal inhabitants and for early settlers. One of the chief means of transportation over these waterways was the dugout canoe. During colonial times these boats were referred to as piraguas by the Spanish and pirogues by the French. The recollections of the early Spanish explorers such as Narvaez, Cabeza de Vaca and De Soto mention dugouts on the Gulf Coast in the early 1500's. . . ."

The article analyzes 4 dugouts, and as a paddler I was interested in their dimensions, keeping in mind that my own boat measured 17 feet long by 3 feet wide with pointed ends. One dugout was 20 feet long and just over a foot and a half wide with blunt ends, one of which had a hole, presumably for a mooring pole. The second canoe had rounded ends and measured 16 by 2½ feet. The third had a pointed bow and square stern, 10 by 1 (!). The fourth featured pointed ends and reached 15 by 1½. I've floated in dugout canoes in tropical countries and found them to be heavy, difficult to maneuver, uncomfortable to sit in, and tippy-feeling, yet they're fast, durable, and get the job done. The most common size I've seen is 36 feet long by 2 feet wide.

Becky grew up in Lucedale and studied biology and anthropology before going to work for the Conservancy. Read said she's the right person for the job: she can back a trailer, run a trotline, or get a truck unstuck as well as any man.

"My dad's biggest love is to go fishing and hunting, and I was a daddy's girl and I would go everywhere with him," Becky said. "Those places just have special memories. And I just remember places like Boneyard Lake and Booger Hole and all these great-named, mysterious places back in the Pascagoula swamp . . . It was a good place to grow up." She also likes the fact that "you can float down it and not see any houses, at least for part of it. You can get back in the woods in the management area and you're just miles and miles away from anywhere."

Becky put some of her memories on paper: "We once owned a camp on Mixon Lake (south of Benndale). To this day, the chorus of certain species of frogs takes me back to summer evenings years ago. I was allowed to use our flatbottom aluminum boat and trolling motor to check limb lines in the lake by myself. There was an air of mystery to the lake; I never could get the boat back into the furthest recesses where there was little open water and lots of big trees. I remember vividly the seed pods of the water tupelo spreading an oily sheen on the water wherever they floated, the sharp crack of a beaver's tail on the water, and I'll never forget how the dry moss on the swollen trunks felt when I pushed against it to maneuver the small boat in close quarters. When you spend much of your time in one of these lakes, landmarks begin to appear—a large knothole, an oddly shaped branch, an arrangement of cypress knees, or a cooler lid lodged in a tree during high water. These things would have been all but indistinguishable from the surrounding landscape by others, but to me they stood out like street signs. Those were some of the best days of my life."

Becky preferred the back swamps to the main river, but I like both equally. For experiencing the whole basin there's nothing like a downriver trip. That way you see the ecosystem unfold, from small creek to wide waterway. Besides, right now the water was too low to explore many of those back swamps. But Scott and I have done so in the past and found much to see, especially in a silent canoe. In a deep, dark, narrow, currentless bayou, the tension mounts as you round a bend, knowing you could surprise anything—like the wild hogs we encountered on a 1992 Chickasawhay float. Four of the critters, young and black, were rooting nearsightedly under the trees, little pigs all in a row making a sound like this: *Uh*! *Oooh*! *Uh! Oooh*! *Uh! Oooh*! I don't know why, but I had the impression the first and third piglets were going *Uh*! while the second and fourth were going *Oooh*! When they saw us they scampered off, but not like we were the big bad wolf or anything. Later, as we headed back down the bayou, the four musketeers burst through the bushes in a major hurry to take care of business. Back on the river, a sow and piglet hightailed it for the woods, panicking up a whole herd of wild hogs.

Becky's job includes managing 5,000 acres of raw woods where such animals live, no easy task. "About a month ago we planted 70,000 hardwoods on Deaton Preserve, and it only rained once. Plus the hogs root them up," she lamented. "We've got a lot of invasive species—cogongrass, privet, Japanese climbing fern, popcorn trees, chinaberry."

I told her I'd seen popcorn trees, aka Chinese tallow trees, in the front yard of her office. She admitted the property also had a couple other invasive species as well. "You can't stop it. You can only suppress it," she said.

We all loaded into her truck and set out for a tour of the preserves. First we went to 1,692-acre Herman Murrah Preserve across the Pascagoula River to the west. Entering the preserve via a locked gate, we drove down a gravel lane through cutover returning to forest. We parked and walked through the woods to a cypress swamp known as Star Lake. It was picture-perfect with cypress knees rising from the water beneath tall trees. The preserve, which includes 3 miles of river frontage, was logged prior to purchase but still serves an important environmental purpose while it returns to full growth. "We're getting all these birds migrating through here like warblers. This is like a five-star hotel for them," Becky said.

A bear was spotted four months ago. There are bobcats, fox, deer, turkey, alligator, and too many wild hogs. Becky even saw a panther years ago in the area, she said. The Conservancy leases the land to a hunting club but remains protective of the wildlife, instructing hunters not even to kill a snake.

We rode to the 3,280-acre Charles M. Deaton Preserve, which has frontage along the Leaf, Chickasawhay, and Pascagoula rivers. Again a locked gate gave way to a gravel lane, this one through more mature-looking forest. We emerged at a natural live oak meadow known as Hog Island, dotted with prickly pear and pocked with gopher tortoise burrows. The tortoises live across the Southeast and are in decline, listed as an endangered species in Mississippi and a threatened species nationally. They live in underground burrows up to 40 feet long and 10 feet deep. Loss of habitat is the main threat, as development and mining operations destroy the places where they live.

Beyond the meadow, the ground sloped down to another picturesque cypress swamp. "We call this the 'ooh-aah' place," Becky said, referring not to wild pigs but to the reaction of students and other groups that visit.

Scott, Travis, and I picked our way into the swamp, marveling at the many-shaped cypress knees, some of which looked like gnomes with faces. "They probably come alive at night," Scott said.

It was nearing noon, so we drove over to the Landmark Cafe in Lucedale for lunch. Owner Keith Tanner, a native of Merrill, joked with us about his backwoods upbringing. "I went barefoot till I was 23. That's when I got my first shoes," he told us from beneath a 10-gallon black felt hat. He sure served good po-boys.

Back at Merrill, I was walking around gathering up my gear when I saw a small cottonmouth snake on the ground beside the camp, apparently headed toward a pile of lumber on the slab under the camphouse. Poisonous snakes are one of the prime dangers on canoe trips, not because they're aggressive but because of the chance of inadvertently stepping on or putting your hand on one. I carry a snakebite kit—the kind with a syringe tipped with suction cups to suck the venom out without cutting the skin—and I've even carried antivenin in Central America. I can't count the close calls I've had—almost treading barefoot on a copperhead, grabbing a log to pull myself up a bank only to see a cottonmouth coiled beside it. It's a testimony to snakes' benign natures that I haven't been bitten. But they're so well camouflaged that the danger is real. Even so, I respect snakes and never bother them in their natural habitat. Only when I see a poisonous variety near my house—or in this instance, someone else's—do I kill them, and then regretfully. The thought that Ransom Jones's children or grandchildren might be playing around this wood pile tomorrow and get bitten removed any hesitation. With no disrespect to the Nature Conservancy, I took my machete and killed the cottonmouth, tossing it down the bank toward the river.

I didn't tell Becky, and I regretted informing Travis when I saw the indignation with which he reacted. Travis objected to killing snakes at all, even rattlesnakes around the yard. "I generally try to catch them and move them," he said. That sounds like a good way to get snakebit, I replied, adding that I have grandkids who romp in my yard and woods, not to mention my plant-loving wife, and can't risk letting a venomous snake range free in that area.

Most people, of course, go too far in that direction, killing any snake they see. But the reptiles—most of which are nonpoisonous—fill an important role in nature, and many species are declining, mainly due to loss of habitat. In particular, the conversion of small farms into pine plantations, subdivisions, and strip malls has robbed many snakes of prey and hunting territory. Fire ants, which are not native to Mississippi, have taken a toll on reptiles by eating eggs and hatchlings. Among species in decline is the coachwhip snake, which grows to 8 feet long, is legendary for its aggressive nature, and likes clear open areas such as broomsage fields for hunting birds and rodents. The glass snake, actually a legless lizard, prefers sandy open soil and is nowadays mostly seen in coastal counties in people's yards. The southern hognose snake thrived in coastal live oak forests, preying on the oak toad, but when the oak forests disappear, so do the toad and the snake.

The black pine snake likes longleaf pine woods, which have been largely replaced by faster-growing loblolly pines. Indigo snakes, which are virtually extinct, were even bigger than coachwhips, but unlike coachwhips they were docile and so were captured for pets. Both the scarlet king snake and the scarlet snake have dwindled since they need hardwood forest, much of which has been replaced by pine. Coral snakes are disappearing, especially from southern counties. Eastern diamondback rattlesnakes are declining due to habitat destruction and wanton killing by people.

We were packing up when Walt and Jo Grayson arrived. They loaded up in the Nature Conservancy's skiff, along with Read and Becky, and followed Travis, Scott, and me as we paddled back up to the juncture of the rivers for more footage. The situation turned comical as we all pointed cameras at each other: Walt wielding his video equipment and nearly everybody else snapping pictures with still cameras.

Then we turned and headed downriver. The motorboat followed briefly before turning back, and we were on our own. Scott, Travis, and I paddled side by side happily, with nothing but a long, wide river and several days of sheer freedom ahead. Travis's GPS put our speed at 3½ to 4 miles per hour.

We stopped at 4 p.m. on a long sandbar on the west bank. There were no tire tracks or even footprints, nothing but water, sand, and woods. Travis found *Listen to the Eagle* on his transistor radio, and I called in to the talk show on my cell phone as he and Scott listened. There was a brief delay, so the words I uttered on the phone emerged from the radio 10 seconds later. The whole situation struck me as bizarre. When I finished talking, Scott went on the air and told his story.

Still full from our big lunch, we snacked for supper and didn't worry with a fire, just stretching out on the soft sand, sipping rum and talking enthusiastically. It seemed we all needed some conversation, especially Scott. "I've got a slight sore throat from talking so much," he admitted after a while. "I've been out there 10 days with no one to talk to."

The topic turned to future paddling trips, and Scott remarked that he'd pretty much seen all the rivers he'd wanted to see. "Heard that before?" I asked Travis, glad to have support for my jaded position.

Scott's interests had transferred to sailing, while my focus was more on gardening, which for me included tending vegetables, fruit trees, muscadine vines, ornamental plants, even a fish pond—anything to do with getting out and working with the land. Still, we admitted to having some interest in other rivers. Scott mentioned having seen aerial photos of Alabama's Tombigbee River basin that made the Pascagoula look like a park—which I took as one of his typical exaggerations. I fantasized aloud about canoeing Louisiana's Atchafalaya River from beginning to end. Travis wanted to do Mississippi's Big Black, a long and respectable river in its own right.

The afternoon was cool and gray, with clouds sliding from west to east. The sandbar tapered down to mudflats and indentations, while upstream, curving white clouds mirrored the river's bend. With dusk, clouds thinned and stars came out. Our enjoyment of the sky turned to intrigue as a jet darted back and forth, then seemed to drop another small plane which in turn released flares. I suggested World War III had broken out, and we pondered what areas we had seen so far would make good hideouts. Then again, it could have been UFOs: people still talk about the pair of Pascagoula fishermen who said they were abducted by aliens in 1973. Scott and Travis, however, showed a typical lack of imagination by pointing out the proximity of Keesler Air Force Base on the coast and concluding we were seeing military maneuvers. No bombs fell or aliens appeared, so maybe they were right.

At dawn Scott made pancakes while I brewed Lapsang souchong tea, a Russian variety with a distinctly smoky flavor. Such nuances were lost on Scott, who still affects a manly taste for black coffee, cigars, and straight liquor. Before the trip I had generously given Scott samples of fine teas and even a tea ball, since he didn't own one, but I wasn't surprised to learn he never tried them. Someday he'll try tea and act like he discovered it. I appreciate good coffee, but too much can scramble your thoughts, which may help explain the previous section of this book.

While we sipped our respective drinks, a turkey hunter eased by in a motorboat through thick mist. He tied up and entered the woods across the river, and we heard a rather awkward-sounding owl hoot, then some yelps as he tried to get a turkey to gobble.

It took Scott a good half hour longer to pack than Travis and me, thanks to the sea kayak's restrictive design. With tiny, cone-shaped compartments, the skinny little boat requires items to be crammed in individually rather than stuffed in convenient duffel bags. With typical maturity and patience, I refrained from commenting at length on his inferior boat, merely pacing back and forth snorting from time to time.

The fog was lifting and the morning felt pleasantly warm when we finally set out. A water snake swam the river ahead of our boats. An osprey whistled contentedly in flight. Black buzzards swooped past, and one soared in tight circles beside me, then flew right over. It cocked its head to look down at me just as I raised my head to look up at it, a rare moment of human-buzzard interaction.

A dried-up bayou entered the river on our right, and we walked up the muddy sand bed among pools of oily-looking water. During high water you would be able to paddle up here. I picked my way through the woods to a bluff where vine-draped trees framed the wide, slow Pascagoula River. Churned-up ground indicated that wild hogs had enjoyed this secluded breezy overlook as well. I joined Scott and Travis in the sandy bayou bed. We reached posted signs, the limits of the wildlife management area. I detoured into a cane thicket, watching for snakes, and made my way to the bayou's source: a muddy, stagnant lake, humming with mosquitoes.

Back on the river we floated past a big cabin tent on a sandbar, probably the camp of a turkey hunter or fisherman. We turned up a channel that led to an oxbow lake, meeting a pair of tanned, impassive young fishermen in a motorboat. Large oxbows are beautiful but essentially dull, at least to a paddler, since there are few coves or channels to add interest. We did sight a musk turtle and, as we returned to the river, an eastern ribbon snake.

At a sandbar for lunch, I took notes, enjoying the scent of Scott's "Irish cream" cigar, then napped in the shade. By the time we left, the headwind had picked up full bore, making the river choppy and mean. Spooked, I paused to tie my bags to the thwarts and zip up my life vest. The last thing I wanted to do was capsize out here. My old canoe creaked and popped ominously as I paddled. We struggled under the Highway 26 bridge, where the wind shoved waves against massive concrete pilings. Scott and I crossed to the right bank where there was a little lee. The wind made me wish aloud that we had opted to use motorboats on this part of the trip. We could have paddled to Merrill, then switched to my 14-foot aluminum johnboat and a 12-foot wooden one Scott built for his father. We could have puttered the rest of the way, unfazed by wind and lack of current.

"Then we couldn't have called it *Paddling the Pascagoula*," Scott said, referring to the title of this book.

"We could call it *Exploring the Pascagoula,*" I whined. "It's just a matter of semantics."

Armed with a double-bladed paddle in his low-lying sea kayak, Scott was unconcerned with headwinds. At last he had found an advantage to his beloved vessel. "It goes under the wind," he bragged. My higher-riding canoe was a real chore to force into the wind, which boomed off the Gulf of Mexico 70-odd miles to our south. The canoe showed its worth when, at the base of a cliff, Scott found an anvil-sized piece of petrified wood and asked me to carry it, my boat being the only one capacious enough for such a load. I protested mightily, but actually the rock-heavy object fit perfectly in the bow and gave me good ballast against the wind. Not that I would tell him that.

At a sandbar in the late afternoon we hiked through a cottonwood grove toward a backwater. But the ground got soft and the deep weeds looked snaky, so we soon turned back. Scott and Travis continued paddling downriver, while I pointed my canoe up to a small channel that led to the backwater. The channel was eerie—narrow, muddy, mosquitoey, thick with the smell of the willows that arched overhead like a tunnel. It broadened to a shallow, mud-bottomed lake with enough coves and branches to make me check my map and compass closely, not to mention look around for alligators. Another slough exited at the far end, and I paddled into a place the map identified as Dead Green River. A Louisiana heron flew out of a cove, squawking. Dead Green River was clearer, with banks sloping up into woods, including some pine trees. The scene felt exquisitely remote, at least until I saw a pair of fishermen in a boat at the far end and realized I wasn't the only human out here.

I rejoined my pals on the river and we made camp on a sandbar.

Scott, in his swift kayak, often has to wait for the slower canoes to catch up.

16

The first bird called at 4:45 a.m., tentatively. Another sounded 20 minutes later, followed by distant wild turkey gobbles. I clambered out of my tent in the dim light, sore all over, hands stiff and swollen from days of paddling. There was no mist, and the dew lay much lighter than it had on other mornings. My fellow adventurers' tents clustered on the sandbar like explorers' shelters on an Antarctic snow field. I lit my stove, its blue flame mirrored in a swath of blue-gray among clouds in the eastern sky.

Somewhere in this vicinity French explorers visited a Pascagoula Indian village in August 1700. In 1699 Pierre Le Moyne, Sieur d'Iberville had arrived on the Mississippi Gulf Coast and settled in at Biloxi Bay. A year later one of his men named Pénicaut went up the river and stopped at a village some 60 miles from the coast. The inhabitants were hospitable if primitive. Men and boys wore no clothing, and females wore only a bit of moss. They lived in cabins with roofs of bark or palmetto thatch and slept on beds made of plaited cane covered with buffalo skins. Their diet consisted of bison, bear, deer, peaches, plums, watermelons, pumpkins, and a porridge made of corn and beans.

Even in 1650 the combined population of area Indian tribes—Pascagoula, Biloxi, and Moctobi–was estimated at a mere 1,000. Numbers dwindled with European contact, and by 1764 the remaining Pascagoula Indians had moved to Louisiana. Britain secured title to the region in 1765, Spain took it in 1779, and the United States annexed it in 1810, adding it to the Mississippi Territory in 1812. In 1817 the territory became a state, and the legislature soon began to improve navigation on the Pascagoula, Leaf, and Chickasawhay rivers, which were already seeing use by cotton barges and other vessels. The boat traffic we'd been seeing was nothing new.

Scott and Travis emerged and we made tea and coffee. Two motorboats came by in the predawn dimness, only one of them with lights: red to port, green to starboard, white at stern. "That's how you can tell which direction a boat is going," explained Scott, our veteran mariner. We ate breakfast while clouds in the west flared pink and subsided to gray as they drifted slowly south.

The water lay still as a lake as we set out down the river. A road ran along the east bank, and we could see cars passing on their way to school or work, a slightly jarring sight. White plastic pipes gushed clear water from artesian springs near camps. Houses on stilts lined the back of a sandbar like condominiums on a Corpus Christi beach. But it wasn't completely tame around here. Travis spotted a 7-foot gator on a sandbar before it fled to the water. A swimming snake paused, head up, to examine my big red canoe. Baby turtles on a log dropped off in a panic when they saw our flotilla. Ospreys, swallow-tailed kites, and American egrets continued to entertain us. We saw our first sea gull, a laughing gull, winging upriver, solitary and black-headed. Soon we saw many more, swooping, landing, bickering, and making their namesake call.

The wind kicked up at 10 a.m., arriving earlier every day. A dirty pickup truck with giant tires was backed up to a sandbar across from a narrow channel which, according to my map, was a back-door entrance to Black Creek, whose actual mouth was miles downriver. We turned up the channel and entered a realm of birdsong amplified by a canopy of trees. Logs had been cut out of the way at the entrance. We got 100 yards before fallen trees blocked our path. An old green Ouachita aluminum johnboat with a trolling motor and short wooden paddle was pulled up to the bank. No doubt the driver of the truck across the river was turkey-hunting in these woods. According to the map, this little bayou joined a maze of others connecting to Black Creek. Pascagoula swamp guide Jay Mengel had told Scott earlier that in high water he sometimes pilots his tour boat through that route. It would make for an intriguing, jungle-like journey.

Back down the Pascagoula, we passed the settlement of Cedar Creek, where rustic houses sprawled under big trees dripping with Spanish moss. Some of the dwellings had decorative touches—windmill, life buoys—plus big pickup trucks with hoods up, dog pens, and lots of johnboats and floating docks. People sitting in the shade asked us the usual questions: Where'd you put in? How far are you going? When we answered, a man warned, "You got a ways to go," not realizing that the remaining miles were a fraction of the distance we'd already come. A cottonmouth snake thick as a hawser flipped off some tires under a bank just beyond the bank sitters, a few feet from where Travis and I paddled.

After lunch on a sandbar, we resigned ourselves to an afternoon fighting headwinds. Both today and yesterday we had to go 2 miles due south, straight into the wind, immediately after lunch. The only respite was sticking close to lee shores, easiest when the river turned east, west, or north. For solace I sang to myself as I struggled: "I Need Thee Every Hour," "Amazing Grace," "Doxology," "Unclouded Day," "I'll Be Your Baby Tonight," "I'm So Lonesome I Could Cry."

Of course, the wind was no trouble for Scott. The only thing that slowed him down was taking photographs of creatures like a yellow-bellied water snake and swallowtail kites. Blooming privet hedge and blackberry bushes brightened the woods, which became gloomy when clouds obscured the sun. In places the river made tight bends with rounded sandbars, the kind you could photograph entirely with a wide-angle lens.

Trying to pace ourselves, we stopped at 3 p.m. on a sandbar across from the mouth of Big Black Creek. This is supposedly a spot where people have heard the legendary "singing river." According to the oft-told tale, the Pascagoula Indian tribe, much diminished and facing extinction by the Biloxi Indians, decided to commit mass suicide by holding hands and wading into the river singing. The chorus supposedly lingers as a haunting reminder. In 1727, the French governor Périer said it sounded "like a swarm of bees in flight." Writing in *Popular Science Monthly* in 1890, Judge Charles E. Chidsey described it as "a roaring, murmuring sound" that rose and fell. A modern-day Mississippian, Edina Rodrigue of McComb, told me, "It sounded more like something tinkling in a glass, but not quite that much glassy sound."

Mrs. Rodrigue, in her 80s, lived in the town of Pascagoula as a child. "Many times my dad would take my two older brothers and myself to the banks of the river to hear the singing river. I heard it many times myself, and it was fascinating," she said. "It was a singing sound, but not singing as such. It was repeated over and over, I guess from the floor of the river." Mrs. Rodrigue said she considered it a natural phenomenon, not Indian ghosts. "I can still hear it. You can absolutely hear it."

Legends about a musical mass suicide existed about the Biloxi Indians, Yazoo Indians, and escaped slaves from Georgia. Even the Pascagoula Indians had such a myth about an ancestral tribe. Such a story is too poetic to dismiss, and the words "Singing River" are pasted onto all sorts of businesses, organizations, and structures in the Gulf Coast area, including the Singing River Alcoholics Anonymous Group, Federal Credit Union, Radiator Service, Cabinets Inc., Radiology Group, Hospital, Glass Company, Rehab and Nursing Home, Fence Company, Yacht Club, Landscape Company, Electric Power Association, Animal Medical Clinic, Industries, Mental Health, Mall, Cinema, Apartments, Mortgage Inc., Baptist Church, Paint and Body, Elementary School, Auto Repair and Uhaul, Grocery, Package Store, Mini Warehouses, Ice Company, Urology, Pharmacy, Blood Center, Ranch, and more. Clearly, this legend is not likely to fade away.

The sun was scorching on the sandbar, but the downstream end rose to form a sandy bluff with a bit of shade at the edge of the woods. Here we found trees for our hammocks, a southerly breeze to keep mosquitoes at bay, and a sweeping view up and down the Pascagoula as well as up Big Black Creek. A few ramshackle camps sheltered in the woods across the river, and some houseboats were tethered up Big Black, but all were vacant on this Thursday afternoon.

It was just a few miles up Big Black Creek to the juncture of Red and Black creeks. The latter is one of the most significant tributaries in the Pascagoula basin, at least from an outdoorsman's perspective. It's canoeable for more than 100 miles, and many people, myself included, consider it the finest canoeing stream in the state. It flows right through the 501,000-acre De Soto National Forest, including the 5,050-acre Black Creek Wilderness Area, as well as the Pascagoula River Wildlife Management Area, and it parallels the state's longest hiking route, the 41-mile Black Creek Trail. Plus it's simply beautiful, with tannin-stained water, big woods, and plenty of small sandbars for camping. Red Creek is a smaller, shorter version of Black, passing through De Soto National Forest and Red Creek and Pascagoula River WMAs, not far from 23-mile Tuxachanie Trail.

The first canoe trip Scott and I took together was on Red Creek, in the late 1980s. We launched during high water and were able to paddle into the flooded woods for a real taste of swamping. That was just a weekend, and it whetted our appetites. A year later we launched sea kayaks at the town of Brooklyn and paddled all the way down Black Creek and the Pascagoula River and out to Round Island, well over 100 miles. In the Mississippi Sound we got into rough conditions, Scott's boat started taking on water, and mine capsized after I selflessly handed him the bailing pump. We made it to Round Island with our lives intact, but that little incident—plus a week of being cramped up in the kayak's little cockpit, enduring its uncomfortable seat, and struggling to coax the slow-to-turn vessel around bends—doused whatever interest I might have had in sea kayaks. It's a testimony to Scott's, shall we say, persistence, that he stuck with them.

Our most unusual trip down Black Creek was when we took a true man of the forest with us. Leslie Minduwa grew up in the Sepik River swamps of Papua New Guinea. I met him in 1985 during my travels in the tropical country, where he worked as an electrician and hunted crocodiles on the side. He, missionary David Sitton, and I spent two weeks traipsing through mountains, hunting jungle birds, and staying in remote villages. Then, in 1991, Leslie reversed the roles and came to the United States. David drove him across the country from Los Angeles to West Palm Beach, Florida, and then back to Mississippi, where he left him with me for a week. David, a native of Texas, had filled Leslie with big expectations for the Lone Star State. "David kept telling me that Texas is the holy land," Leslie told Scott and me in his British-accented English. "When you come to Texas you are in heaven. We drove and drove, and there was just nothing. I said, 'Where is this Texas?' and he said, 'This is it.' I said, 'Man, this isn't heaven, this is hell.'" He laughed. "I think I'll tell my friends to go to Mississippi and leave the rest."

Leslie formed his opinions about Mississippi mainly during our three-day canoe trip down Black Creek. We put in at Brooklyn, and the instant Leslie dipped a paddle into the water I knew I was in the presence of an expert. On the Sepik River, canoeing is the primary mode of travel, and Leslie grew up doing it. Paddling in the stern, he propelled the canoe seemingly at the speed of a 35-horsepower outboard. He gave me lessons, showing me the guide stroke—I got that name from a book on canoeing, incidentally, not from Leslie—which he considered superior to the J-stroke. The J-stroke is the classic technique by which the stern paddler keeps the boat going straight while paddling on one side. If you've ever paddled a canoe from the rear, you know the bow tends to turn in the direction opposite the side you're paddling on. To execute a J-stroke, you flip the back of the paddle blade away from the hull at the end of the stroke to counteract the turn. In a guide stroke, you flip the front of the blade out. One way to illustrate the difference is that when you do the J-stroke, the thumb of your top hand points down; when you do the guide stroke, it cocks back toward you. The J-stroke is smoother but puts more torque on the wrist—significant to a carpal tunnel syndrome sufferer like me—and feels a little less forceful. I've done the guide stroke ever since Leslie showed it to me. Traditional North Woods paddlers like the late great Bill Mason, however, dismiss it as the "goon stroke" because it's not quite as elegant. Sorry, Mr. Bill, but I consider Leslie—a jungle tribesman who was canoeing before he could walk—more authoritative even than Canadians.

I wasn't prepared to adopt all Leslie's habits, though. "On the Sepik, men stand up when they paddle," he told me. "Women sit down." In the larger dugout canoes, as many as 15 men stand, with a man in the stern using a paddle as a rudder. With forceful strokes they easily advance upriver against the current. For the Black Creek trip, Leslie adapted to our wimpy ways, sitting in the back as we floated downstream.

Our first night we cooked sweet potatoes, onions, and plantains (cooking bananas) New Guinea style, burying them in the sand and building a fire over them. In my past travels I had learned to speak New Guinea pidgin, and Leslie and I conversed at length, with Scott trying to follow. The next day Leslie tried my rod and reel, his first time to use such an apparatus. On the Sepik River, men use drop lines, spears, or nets, and at certain times of year, in low water, the fish swarm to the surface by the hundreds and can be scooped into canoes with paddle blades. Here on Black Creek, Leslie quickly caught a pair of pan-sized bass. He scaled and gutted them, leaving the heads on, then salted the inside and lay the fish on the coals with a stick through the lower jaw. After 20 minutes on each side, the skin peeled back to reveal perfectly cooked fillet, which we ate with our fingers. He showed me how to eat the heads and suck the eyeballs, which believe it or not were delicious. Don't feel shocked: many Mississippians gladly suck crawfish heads and eat squirrel brains, hog head cheese, and chitterlings, among other delicacies. His method of cooking the fish was much simpler than our own technique of battering fillets in corn meal and frying in hot oil. I told Leslie, "Americans think we have to have $1,000 worth of equipment to go camping." He laughed. To him, even pots and pans were a luxury. I suspect his camping and canoeing habits weren't all that different from those of the native Americans who lived along the Pascagoula.

The mosquitoes were bothersome to Scott and me, but Leslie hardly noticed them. The Sepik, of course, is notorious for its voracious bugs. Leslie remembered taking an Englishman to his village, and though the man used repellent, the insects attacked him so fiercely that he panicked. "He ran up and down the village, swatting himself," Leslie recalled with a laugh. When I asked Leslie how he deals with mosquitoes, he said, "Just ignore them." He took advantage of our repellent, though.

Crocodiles pose a more serious threat in parts of Papua New Guinea. The saltwater variety grows more than 20 feet long and 4 feet wide, and thinks nothing of preying on humans. Alligators, in contrast, are generally not aggressive toward humans, though they might have been in earlier times when they lived long enough to get really big.

Another danger of the Sepik area is sanguma men, or sorcerers. Sorcery may be alien to our culture now, but it's real in Papua New Guinea, just as it was among native Americans. Several years ago, Leslie's wife, Rose, fell prey to a sorcerer seeking vengeance against her brother. She was walking along a path when she heard a whistle. She turned and saw the sorcerer, who cast her into a spell by making mysterious hand motions. Rose became ill, and later her body went numb. Leslie took her to a hospital, but doctors could find nothing wrong. Devoted Christians, the Minduwas prayed for help. Just then an old man came over. Summing up the situation, he advised Leslie to stick his wife with a pin to see if she were truly bewitched. Leslie slid a pin 2 inches into his wife's forearm; she didn't react. At that the old man made an infusion of medicinal herbs for Rose to drink. It made her violently ill, but cured her. If not for that, Leslie said, the sanguma man likely would have instructed Rose to allow him to inject her with poison, killing her. Leslie is an educated, intelligent man, an elder in his nondenominational church. When I asked him how sorcerers have power, he said simply, "I think it must be something supernatural." I explained that to an American like me it seems more like a case of hypnosis. He shrugged.

In other ways, life in Papua New Guinea is not that different from life in modern-day Mississippi. The people are friendly. They live in small towns or villages, and enjoy hunting, fishing, and gardening. It's not hard to understand why Leslie felt at home here. Yet his visit gave me insight into how native Americans might have lived in the Pascagoula basin.

As Scott and Travis napped and read, I bathed in the river, slung my hammock among the trees, and tried to read. My gaze kept drifting back to the peaceful surroundings: scaly-barked sycamore trees, saw palmetto bushes, yellowtop flowers, oak saplings, blackberry canes. A pileated woodpecker bugled its jungle cry. With rain threatening, Travis and Scott pitched their tents in the woods, while I walked to the crest of the high sandy bluff to erect mine. Other than an occasional drizzle, the rain never materialized, and we ate supper, chatted; then Travis turned in early to read. I played my guitar as a yellow crescent moon settled toward the woods across the river; then I handed the instrument to Scott. As he plucked the strings, an owl—his spirit animal—landed on the beach beside him. It stayed for an instant, then flew. Amazed, I told Scott about my close encounter with a great blue heron on Bogue Homa Creek, and we agreed our spirit animals were looking out for us. That may seem odd coming from a Christian, but owls and herons are God's creatures, too.

After the moon sank into the forest, Scott went to bed, and I walked up the bluff to my tent. The nighttime view was astonishing: starry sky, black woods, glimmering rivers. I said a prayer, thanking God for this beauty, this trip. When I said amen—I'm not kidding—the river began to sing.

The timing was too perfect to be true, but there it was, a high-pitched, eerie sound coming from the dark wall of forest just above the mouth of Big Black Creek on the west bank of the Pascagoula. I stepped to the edge of the bluff and peered, trying to identify the source. I'd been out here close to two weeks and listened to all sorts of night noises, but had heard none quite like this. Nor did I recall hearing it on previous trips. This was closer to Judge Chidsey's "roaring, murmuring sound" than Périer's "bees in flight" or Rodrigue's "tinkling in a glass." If I had to guess, I'd say it was some species of frog, but it seemed odd I hadn't heard it earlier. And why did it start when it did, like a postlude hymn following a benediction?

I couldn't explain it, and didn't need to. It was just one of those magical experiences that occur unexpectedly when you're out in wild places, hinting of connections that are deeper than we can fathom. I felt comforted, not spooked, when I slipped into my tent and went to bed.

Travis Easley investigating a tributary on the Pascagoula. Winding waterways like this lead off the Pascagoula into the thousands of acres of hardwood forests in the river basin.

PART FOUR

LOWER PASCAGOULA RIVER

SCOTT B. WILLIAMS

Sandbars become smaller and scarce in the tidal zone of the lower Pascagoula River. This is our campsite from the last night of the trip, before we began the final push through the marsh to our takeout point in Gautier.

17

Since Ernest has described the upper reaches of the Pascagoula River, as well as my favorite of all streams, Black Creek, I am left to narrate the remainder of our journey to the coast along the lower river. Although I've had enough adventures on Black Creek to fill another book this size, we agreed that Ernest would write about its extraordinary canoeing potential. Black Creek is my favorite of all streams, despite the fact that I've paddled hundreds of waterways in dozens of states and even a few foreign countries. I have yet to find a place offering such consistently perfect conditions for camping as the national forest lands and gorgeous white sandbars flanking Black Creek. Every time I return there I wonder why I even bother to ever paddle anywhere else. I've lost count of the trips I've made down Black Creek's upper stretches. Even here, now that we were camped at its confluence with the Pascagoula, I was in familiar territory. I had passed this way many times, as over the years I had made several trips down the entire length of Black Creek to the Pascagoula River and on to the Gulf.

Ernest mentioned our kayak trip back in 1988. That was a false start to my much longer trip down the Florida coast and on to the Caribbean. A few years later, I paddled solo the length of Black Creek to the Pascagoula, and on to the Gulf, then turned around and paddled upstream the entire length of the Pascagoula and part of the Leaf River to McLain. Going upriver was an experience I don't care to repeat, but at that time I was gung ho for paddling, and in a swift sea kayak, it's entirely possible to make good time traveling upriver, especially if you play the eddies and stay out of the main current as much as possible. If there is one way to get to know a river, it is to paddle it upstream. Traveling at a much slower pace and working for every mile, it's hard to miss any subtle nuances of the river and especially of its currents.

I felt at home the previous afternoon when we agreed to make camp across from the mouth of Black Creek. When a barred owl, the creature that I have long considered to be my spirit animal, flew across the river from the side where Black Creek enters and briefly alighted at my feet that night as I played Ernest's guitar, I took it as a good omen. I felt nothing could stop us now and all that remained was a leisurely paddle until we reached the salt of the Gulf.

As I knew it would be, the experience of traveling in company with Ernest and Travis on the Pascagoula River was totally different from my solo journey down the Chickasawhay. Having spent so many days alone, I was eager for conversation and talked until I was hoarse that first night on a sandbar below Merrill. The only time I get lonely when traveling alone is at night, when the day's progress has been made and camp chores are done. This is the time that camping with companions is most enjoyable, sitting around a fire, reminiscing about other trips or discussing topics such as philosophy, religion, or the mysteries of the universe. Such campfire conversations are, of course, aided and enabled by the passing around of a bottle of whiskey or some other dialogue-provoking convivial beverage, and when camping with others I tend to stay up later as the time passes much more quickly than for the lone traveler.

The disadvantage of traveling in company becomes apparent first thing in the morning, when people can often have quite different ideas about when and how to proceed with the beginning of the next leg of the trip. I already knew what to expect from my many river trips with Ernest. I'm an early riser, up and moving by dawn, both at home and in the outdoors, but once I'm up I like to take my time and ponder the possibilities of the day over at least 2 or 3 cups of coffee before I even eat breakfast. Ernest, on the other hand, as I've noted before, doesn't even appreciate coffee anymore. Even when he did, he would take a sip or two and start packing his sleeping bag and emptying his tent before finishing the first cup. Then, he might take one more drink before dramatically flinging the remaining coffee and grounds out of the cup and shouldering his river bag to begin loading the canoe. It was no different now that he was drinking tea. Each day he was ready to go before Travis and I were, and, having to load my stuff through the small hatches of the kayak, I was the last one on the river every day. Not once did either one of them wait until I was ready to go. I wolfed down my pancakes or oatmeal each morning as Ernest paced back and forth along the water's edge with growing anxiety. It got worse by the day, and as we began stirring that morning at the place where Black Creek joins the Pascagoula, he announced that he was going to launch well before either Travis or I began striking our tents. Travis left about 30 minutes later, and when I finally slid my kayak into the river and climbed in, Ernest had at least an hour's head start. I had no doubts that I would catch up with both of them eventually, but it was aggravating, because being widely separated meant that I could inadvertently pass one or both of them if they turned off to explore a side channel. Likewise, if I saw such a place I wanted to investigate, I would feel rushed because they might be pulled over somewhere to wait for me.

Once they were out of sight though, I forgot about these worries and focused on my surroundings. About three miles below the mouth of Black Creek, a bridge crosses the Pascagoula River at the community known as Wade. There's not much of anything at Wade really, but a boat ramp and a few houseboats tied up along the bank. These houseboats become more common the farther downstream on the river you travel, although there are now far fewer than there were in the days when they were allowed to tie up anywhere along the river. The owners of these semipermanent floating camps were eventually forced to remove them from parts of the river flanked by public lands, so today none are seen in areas encompassed by the Pascagoula Wildlife Management Area.

Most of these floating camps cannot really be described as "boats" by any definition of the word. The majority of them are homemade shacks floating on a collection of empty oil drums, Styrofoam blocks, or cast-off industrial barges. They are not designed to move, and many are even permanently wired to electricity and running water from shore, especially in the more accessible areas like the bridge crossing at Wade.

Passing these camps at the Wade Bridge reminded me of meeting a real Pascagoula River character here when Jeff Hudson and I paddled this way almost 10 years before en route from Hattiesburg to the coast. It was the same trip I already mentioned, which started off with our losing my hand-built wooden canoe the first night out due to rapidly rising water on the Leaf River. After that incident the trip had been mostly uneventful until we reached Wade. As we were paddling past the last of the houseboats, making good time in the swift currents of mid-river, someone practically screamed:

"HEY! Y'all want some fish?"

We looked at each other and did a quick U-turn in the river to paddle back to the houseboat on the east bank where a red-headed man in camouflage and rubber boots was standing and waving from the deck. Fresh fish would be hard to turn down after nearly a week of canned tuna and rice. Seeing our reaction, the man stepped into his johnboat that was tied up to one end of the houseboat and began pulling up something that was obviously heavy.

"Y'all eat gar, don't ya?"

We looked at each other again, realizing we might have made a big mistake in turning around, but as the man pulled up his submerged fish trap he laughed and we could see that it was full of 1-to-2-pound catfish. He introduced himself as DeWayne, and invited us aboard for a cold beer. It was only 9:00 in the morning, but there was something about DeWayne that was hard to refuse. Maybe it was the way he was honing the foot-long butcher knife he continued to work on as he asked us why we were traveling in a canoe, where we had come from and where we were going. A shotgun was propped up against one rickety wall. A couple of worn-out looking hunting dogs dozed on the deck near the johnboat. Something flopped in the sink as we talked and I looked over to see that it was a duck, still barely alive until DeWayne went to work on it with the knife after opening another beer.

"Y'all want some squirrel?" DeWayne asked, pointing to the stove top in the one enclosed room of the floating cabin. How could we say no? We each took a greasy piece of meat out of the iron skillet and ate as DeWayne told us he was staying out here on his riverside camp for the entire fall and winter hunting season. He had a wife in Hattiesburg or somewhere, but she knew that when hunting season rolled around, he was moving to the Pascagoula River. He worked some sort of construction job that allowed him the flexibility to just quit each year in hunting season. Out here, he hunted daily and set traps and dropped lines for catfish. It was obvious DeWayne knew the river and knew how to harvest the fish and game to be found in or near it. He just wanted to talk, and we obliged him until we felt it was safe to ask to be excused; then we paddled on downstream and stopped on a sandbar to fry the catfish for lunch.

I half expected to see DeWayne again as I paddled under the bridge at Wade, but no one was around and I didn't even see a houseboat that looked like his. I caught up with Travis before long and paddled alongside his solo canoe for a few minutes, until it became apparent he wasn't interested in conversation this morning and just wanted to poke along at his own pace. Travis and I had really hit it off since our rendezvous at Merrill, and we spent a lot of time talking about our various adventures and comparing notes on places we had visited. At 29 years of age and free and single, Travis was ripe for more travel and asked me lots of questions about my long journeys. As Ernest noted, Travis was perfectly comfortable out in the bush and I had no doubt he was a good candidate for long, solo adventures of the sort I had done when I was his age. We also had talked at length about sailing, and I entertained hopes that I might have found a potential crew member to help me make a couple of offshore passages I had long dreamed about undertaking in my 26-foot cruiser, *Intensity*.

Since he didn't want to talk about sailing this morning, I paddled on ahead, still seeing no sign of Ernest, who was obviously making miles and probably hadn't stopped to take a break all morning. We had been encountering strong headwinds each afternoon, making progress in the canoes difficult, and Ernest wanted to get as many miles as possible behind him before the wind started blowing.

I passed a couple of johnboats, one pulled up on a wooded bank with the owner nowhere in sight. It was turkey season, and many of the river users we'd seen on the Pascagoula were turkey hunters. Just as I passed this boat I heard a gobble from the woods nearby. I couldn't tell if it was the hunter's call or the real thing, but shortly after I rounded the next bend, I heard the report of a shotgun. I stopped on a sandbar and watched as several swallow-tailed kites circled overhead and swooped in low graceful arcs almost to the surface of the river. When Travis caught up, he said he had met the turkey hunter, who indeed had a gobbler in his johnboat and had stopped to show it off and chat for a minute.

Eleven miles below our start point, we at last caught up with Ernest and found a place to have lunch just across the river from the entrance to a large oxbow lake. Such lakes become more common on the lower reaches of the river, and in times of high water it is possible to paddle for miles off the main river following enchanting channels of cypress-lined, still waters. We did just that the first time we canoed a stretch of the Pascagoula many years before. The river was far out of its banks and with a canoe it was possible to paddle just about anywhere through the surrounding forest.

On that trip we made several excursions off the river. Ernest sat in the bow clearing a path through the bay thickets with a machete while I paddled and pushed from the stern. We followed a compass course deep into places we never would have seen otherwise. After one whack at an overhanging limb, a small water snake, thankfully the nonpoisonous kind, fell into the canoe. I flipped it overboard with the end of the paddle. We discovered a monster of a hollow cypress tree, with a trunk so wide I could stand in the opening at its base like it was the mouth of a cave.

With the water levels unusually low for this time of year, such off-river paddling was out of the question as Ernest and Travis and I made our descent of the river. Since the lake across from our lunch stop was occupied by several houseboats and a couple of johnboats buzzing back and forth from the river, we made only a tentative foray into its entrance to shoot some photos and then continued on downriver.

As expected, the south wind picked up shortly, extracting a severe penance from Ernest and Travis, who chose to travel in such high-profile boats with nothing but a single-blade paddle for propulsion. As a sea kayaker, I enjoyed a bit of headwind for its cooling effect and the feeling that I was on big water where my boat comes into its own. I had no trouble maintaining my usual pace, despite the wind, so I often pulled far ahead and then waited in a sheltered bend or on a sandbar for my partners to catch up. The Pascagoula is much bigger below the confluence of Black Creek, so the bends are farther apart. Long, straight stretches of north-south facing reaches make it possible for the wind to kick up quite a chop of small whitecaps, making life even more difficult for canoers. Knowing they would need a break, I ducked into an interesting-looking side creek entering the river from the west bank and waited. When Ernest and Travis finally got there, their attention was so focused on forward motion and fighting the wind that they didn't even see me until I called out to them.

The side creek was narrow, barely wide enough for two canoes to pass. The banks were steep and made up of slippery mud. An ancient cypress tree several feet in diameter at its base blocked any possibility of exploring upstream in the boats, but when Ernest and Travis pulled in I suggested a short foray on foot to see what we could find. The forest here was typical of the bottomlands that flank the lower Pascagoula. Towering hardwoods practically shut out the sunlight, and the understory was thick with palmettos, vines, lush ferns, and wild lilies. We left the boats and carefully followed the bank, keeping an eye out for hidden snakes, and marveling at the size of the trees. Not far from the river, we came to a cypress-filled dead lake about a hundred feet north of the little stream. Wallowed-out mud and slide marks indicated an obvious alligator highway that connected these two hidden backwaters, but though we approached in silence, we didn't see any of the big reptiles. Travis shot some footage of the surrounding forest with Eddie McCalip's camera; then we made our way back to the boats and he broke out his stove to make some espresso.

To my surprise, even Ernest accepted a cup of this afternoon caffeine treat, and we enjoyed a half hour break from the wind as we drank and discussed possibilities for the evening's campsite. Another small community called Cumbest Bluff was not far downstream, so we had to get past that to find a place sufficiently remote for camping.

We soon reached Cumbest Bluff, and found it to be a surprisingly big community for such an out-of-the-way location on the banks of the Pascagoula River. Occupying maybe a half mile of riverfront on the east bank, this collection of houses appears almost suburban compared to the ramshackle camps we'd seen on other parts of the river. Most of the houses here are permanent dwellings, and many are complete with grassy lawns and private sand beaches on the waterfront. Posted signs were everywhere on these sandbar beaches, so there was no temptation to stop or even slow down as we kept to the strong current of mid-river and maintained a steady pace.

Below Cumbest Bluff, the river became wild again, but we knew from the maps that it was not far to the place where the river splits into the East Pascagoula and the West Pascagoula. We wanted to camp upstream of this split, as we had already decided we would all stick together and paddle the west river on to Gautier. Our original plan had been to split up again, with me taking the east river in the kayak, then trying to arrange another rendezvous at Gautier. But logistics made it simpler to stay together. We would reach the coast in one more day, and be dependent upon Eddie McCalip, someone in Travis's family, or Ernest's wife, Angelyn, to pick us up at our takeout point.

Just below the split on the west river was another riverside community called Poticaw Landing, and below that sandbars become scarce to nonexistent. We studied the maps of the area below Cumbest Bluff and determined that there was a good sandbar for a campsite before the split but below a place where a high-voltage transmission line crossed the river. We reached this sandbar at about 4:00 p.m., and found that it offered a narrow beach for pitching the tents, but no shelter from the sun for at least two more hours. We pitched Ernest's huge four-man tent and sat on the shady side, drinking the last of the Captain Morgan while waiting on cooler evening temperatures before cooking dinner. Travis cooked some tasteless concoction of freeze-dried mashed potatoes and English peas, and we rounded out the main course with nuts, raisins, and other lunch-time snacks.

The tiny sandbar was high and dry, but provided only a narrow buffer between the river and the jungle-like vegetation that started just beyond the sand. Travis had a close encounter with a cottonmouth while walking through the cattails at the end of the bar beyond where we had pulled up the boats. I pointed out to Ernest and Travis that we were now in the tidal zone of Pascagoula estuary, and proved it by planting sticks in the sand at the water's edge so we could monitor the ebb or flow and determine whether it would be rising or falling during the night. I later regretted this as it sparked a whole new sense of urgency in Ernest, who concluded from the experiment that the tide was now falling and we would definitely have to get up well before daylight and paddle like crazy for hours in order to beat the next incoming tide. It was as if he expected a raging tidal bore that would rush upriver from the Gulf and flush us clear back to Merrill if we did not get up and make 15 or more miles before mid-morning.

Travis, on the other hand, was fascinated by the connotations and possibilities of the tidal zone. As I pointed out to him, this simple rise and fall of water implied a connectivity with all the waters of the world, and from here, given enough time, one could paddle *anywhere.* He liked that idea, and once again I was able to go off on a tangent about the appeal of sailing, how this was the reason for owning a boat like *Intensity*, because most of the earth is covered in water. Ernest wasn't interested. He moved the little sticks a bit farther and studied the water to see what was happening. Nothing would convince him that there was any option but an all-out race against the tide. I crawled into my tent and went to bed. I knew exactly what to expect. We would bust butt all morning and then have hours of waiting when we got to Gautier. Just like Brus Laguna.

Wild spider lilies, several inches across, are common in the coastal marshes of the river delta.

18

I woke that last morning of the trip to the sound of Ernest's rubber boots squeaking in the deep sand and his rummaging about in his packs for his stove so that he could make morning tea. It was still pitch black outside. I looked at my watch. It was just a few minutes after 5:00 a.m. Ernest was unstoppable. I pulled on my T-shirt and stumbled out in the dark with my Mini Maglite to start my coffee. I was determined to have a proper breakfast and to take a bath in the river before leaving. After all, we would be arriving in civilization today. I still had clean clothes packed away deep inside my kayak just for this occasion. Ernest and Travis could look like river rats if they wanted to, but I considered the ability to come out of the woods after two weeks freshly bathed and wearing clean clothes to be the ultimate test of woodcraft.

I worked quickly to get things done, because I didn't want us to leave at separate times like we had the day before. This section of the river on down to the marsh presents a maze of options, because of river splits, dead lakes, and shortcut bayous. It would make a lot more sense to travel together. To speed things up, I made oatmeal instead of pancakes, and while Ernest and Travis were tying their bags down in the canoes, I walked around the bend upriver and dove into the cool water for my bath. We were under way by 7:00 a.m., our earliest start as a group since we had met at Merrill. It was still dead calm, and the tide was slack. Progress was easy as we ghosted past jungle-like stands of hardwood forest where the trees overhung the edge of the river and sandbars were nonexistent. Birdcalls echoed out of the stillness, punctuated by the occasional splash as a mullet broke the surface of the river. We were obviously close to the Gulf.

Huge palmettos crowded each other for space at the narrow sunlit edge of the forest along the waterfront. Many of these had trunks extending out of the ground like the wild cabbage palms common in the woods of Florida. They added to the jungle effect, as did large patches of wild lilies, tall ferns, and tangled vines. This part of the lower Pascagoula is noted by botanists for its plant diversity, with higher numbers of species per acre than practically anywhere else in the state. Wild orchids even grow here, 21 varieties in fact, including the Green Fly Orchid, a true epiphytic species like those found in tropical rain forests that cling to overhead branches in the trees.

The community of Poticaw Landing is located on the west side of the West Pascagoula River just downstream of the place where the river splits. Unlike Cumbest Bluff, Poticaw Landing is made up mostly of houseboats, many of which have been there for decades and reflect the character of their owners, who have probably spent most of their lives in intimate knowledge of this part of the river. Unlike some other man-made intrusions we passed on the river, Poticaw Landing seems to belong there, and passing it reminded me of paddling past native villages on the banks of the Patuca River in Honduras. Some of the residents were out on the porches of their floating homes, fishing or just enjoying the morning, and we had passing conversations about our trip as we drifted by. Most of the people we encountered in Poticaw knew what we were up to, thanks to all the publicity we had received in the *Mississippi Press* and on radio and TV.

Below Poticaw, the west river is narrow and winding. We passed another community at Roger's Bend. A possible shortcut shown on our dated maps proved to be a dead end. The river had changed course since the maps were printed, leaving the lake we hoped to cut through filled in with willows and other swamp vegetation. Many of the old oxbows in the area are like that.

We found one last sandbar in an area where the banks were nothing but forests or cattail marsh, and stopped for a final opportunity to stretch and eat a snack before reaching the true coastal marsh where we knew we would find no dry land until we got to our takeout at Gautier.

While kayaking I like to keep a small 10-liter dry bag in the cockpit with me for just such occasions where I can't stop or get out of the boat to open the hatches to the storage compartments. This is especially important on open water crossings while sea kayaking, so I have always been in the habit of packing this dry bag with high-energy snack foods, sunscreen, insect repellent, drinking water, and other essentials before leaving shore. When I'm alone, I have a system that is second nature from long habit. Everything has a place and everything is in its place. Such systems go out the window when traveling with others. Conversation and the general distraction of the others' activities while breaking camp can make it easy to forget something.

To my dismay, when I opened this dry bag after we pulled up on this last sandbar, I discovered that what I had forgotten was to tightly close the lid on the one-liter Nalgene water bottle I put in there before we left. The bag was now full of water. Raisins, granola bars, and Wheat Thins were afloat. Deeper in the bag I had stuffed my rain jacket, my .45 automatic pistol and its ammo, a compass, and my Epipen injection for bee or wasp stings. Thankfully no cameras or electronic gear went into the bag. I dumped out the water and carefully wiped down the gun. I told Travis that if Ernest hadn't been so frantic to leave to beat the nonexistent incoming tide, I wouldn't have had to pack my bags in the dark.

The south wind had already begun to blow, however, so I knew it had been a good decision to leave early. There were big stretches of south-facing open water ahead, and Ernest and Travis were in for a battle. We left the sandbar and paddled on, winding ever downriver in a seemingly endless succession of big, looping bends. We finally reached Swift Bayou, a shortcut Ernest and I had discovered our first time down the Pascagoula River. Swift Bayou leaves the West Pascagoula well north of Interstate 10, providing a twisting but narrow and protected route through the marsh to the point where it rejoins the river well south of the twin spans of the bridge. I knew it well as I had used this route every time I paddled the river. I had been warning Ernest and Travis all along about the monster alligator that I knew to be a resident of the bayou, but since they hadn't seen this creature first-hand, they could not be convinced and seemed doubtful when I tried to convey the size of this great reptile. Jeff Hudson and I saw it when we passed this way on the same trip in which we had lost the canoe and met DeWayne, the hunter and trapper from the houseboat at Wade. I had spotted it first, stretched out in the marsh grass on the edge of the bayou. From a distance, it looked like a pile of old automobile tires, but as we paddled closer, I could begin to make out a frightening form. We didn't think it could be real, it was so big, like a plastic mock-up for a movie set such as the shark used to film *Jaws*.

Covered in slime and bits of grass that hung from the dinosaur-like scales of its back, it looked like a creature from another world as it lay motionless on the bank. Swift Bayou is aptly named for its current, and Jeff and I were about halfway down it when we saw this alligator. The only option to avoid passing within 20 feet of it was to paddle all the way back up to the West Pascagoula and take the long way around to Gautier. We were tired and ready for our 8-day trip to end.

Careful not to dip a paddle, we drifted until we were a little past the sleeping gator. It was real, all right, and as big as it had looked from a distance. It had to be near the upper limit of its species size potential. I knew there were records of alligators over 18 feet. In the frontier days such gargantuan sizes were not at all uncommon. These days, even a 12-to-14 footer is considered huge and relatively rare. To be sure it was alive, as we were just past it I slapped the water hard with my paddle blade to see if I could get a reaction. We both felt the chill of fear as the massive head lifted, eyes opened, and the lower jaw hung agape to reveal a nasty row of wicked teeth. To my way of thinking, there's not much difference between an alligator this size and a Nile or saltwater crocodile. I think part of the reason alligator attacks against humans are so rare is simply because gators big enough to think of people as food are few and far between. This fellow could have eaten a couple of grown men like Jeff and me with no problem. And I knew that despite the sluggish appearance, it could surge to life with lightning speed and catch us if it wanted to. Needless to say, we dug in with the paddles and put Swift Bayou behind us as fast as possible.

Before I turned into Swift Bayou with Ernest and Travis, I could hear distant traffic from Interstate 10 and knew that for the first time since crossing Highway 98 on the Chickasawhay River, I would have a working signal for my cell phone. I called Michelle to let her know we were alive, enduring Ernest's remarks of ridicule as he took a photo of me drifting in the kayak with the phone pressed to my ear. We turned into Swift Bayou after I put the phone away, and I climbed out on a log, the only solid place for miles, so I could relieve myself one last time before we entered the marsh, where getting out of the kayak would be impossible. As I stood there while Ernest and Travis drifted, we heard the roar of an outboard and I slipped into my kayak just in time as a large, canopied tour boat full of blue-haired old ladies sped around the bend from downstream on Swift Bayou. The tour operator slowed down and waved and several of the passengers snapped pictures or shot video of us; then they were gone. In Ernest's estimation, it was a close call. It would have been far more damaging to my adventurous kayaker reputation if they had seen me adrift, talking on the cell phone, than if they had rounded the bend and caught me in the embarrassing position they had missed by seconds.

Though I peered hopefully ahead into the grassy marsh as we rounded each bend, there was no sign of an alligator, giant or otherwise. The noisy tour boat had likely sabotaged our chances of seeing one, but Ernest and Travis were now smugly convinced that my sighting with Jeff was just another tall tale. The bayou was alive with other life, though, and frequent splashes and ripples in the water ahead indicated some kind of big fish fleeing at our approach. Redwing blackbirds were everywhere. Wildflowers were abundant too, most notably Louisiana iris, white morning glory, and weird-looking white spider lilies several inches across. This coastal marsh environment was new to Travis, and I couldn't tell whether he liked it or not. He seemed a little uncomfortable with it, being a woodsman at heart. It is certainly disorienting for anyone new to it, especially traveling low to the water in a canoe or kayak. The tall grasses and reeds that make up these expanses of marsh are impossible to see over, and big rivers like the Pascagoula diverge into many channels in the last few miles before entering the sea, creating a maze for navigators. The good thing, though, is that by following the current downstream, you are sure to get out; it just might take longer than expected. The marshes of the lower Pascagoula are unfortunately not as easy to get lost in as I would like, thanks to man-made landmarks such as bridges.

We passed under the noisy spans of Interstate 10's west- and eastbound lanes, and as we entered more open water downstream of the bridge I looked up to see that the only drivers who could see us over the high guardrail were the operators of high-riding 18-wheelers. Most of those above us were unaware of our passing, and undoubtedly wouldn't be able to comprehend where we had come from and what we had been doing for the past 2 weeks. Below the bridges, Ernest and Travis were forced to stay close to the tall marsh grass on the east side of the channel in an effort to get out of the direct force of the wind. I pulled on until I was far ahead, and then nosed the kayak up into the reeds to rest and wait for them to catch up. This was the pattern for the rest of the trip. It would have been much more effort for me to slow down to try and match their pace, so I punched on across each open stretch and then played the waiting game. Now more than ever, my boat was in its element, as each body of open water got larger, with corresponding waves to match. I felt sorry for my companions, but realized that they were lucky too. It could be much worse. The wind was blowing maybe 12 to 15 knots. If it was 15 to 20 or more, the waves would be more than they could handle in the open canoes. It's not uncommon to have 3-to-4-foot chop in the wide-open stretches of the lower river.

The tour boat passed us again, speeding back downriver, and then about 20 minutes later we saw it coming back, this time empty except for the crew operating it. They were the McCoy brothers, owners and operators of McCoy's River and Marsh Tours, and they wanted to know all about our trip. I pulled my kayak up alongside them and clung to their rail while I gave them the short narrative. Ernest and Travis paddled on. They were miserable in these conditions and didn't need any distractions. For them, getting to Gautier would require single-minded focus on putting one paddle stroke after the next.

I exchanged phone numbers with the McCoy brothers and set out to catch up. The wind had increased a lot by this time. It was nearly noon and it would get even worse later. We passed a fisherman who was having good luck with red drum, saying he had caught several despite the windy conditions. Once again I left my companions far in my wake and approached Gautier alone, rounding the point at the entrance to Mary Walker Bayou and spotting Tucie's Fish Camp, our takeout point, dead ahead and welcoming after the long slog to windward. It was 1:20 in the afternoon when I got out of the kayak at the dock there and explained the trip and my boat to two curious fishermen. Ernest and Travis arrived 25 minutes later, and we found a place beside the boat ramp to haul our boats out on the grass. They were beat from fighting the wind. My first priority was to find a place to change into the clean and dry clothes that I had saved just for this occasion. Ernest and Travis could have cared less about that, but then they weren't nearly as wet as I was, riding high and dry in their canoes while I had been sitting in a puddle of water all day.

We sprawled out on a grassy hillside overlooking our boats and Mary Walker Bayou and ate a lunch of cheese and crackers. We had an appointment for an interview with Brad Cockburn, the reporter for the *Mississippi Press* who had telephoned Ernest and me before the trip to get our story. He arrived within an hour with a photographer and a list of the usual questions, and we posed in front of the canoes and did our best to describe what the experience had been like. Read and Becky Stowe, who had been so helpful during our stopover at Merrill, arrived a little later that afternoon, and we went with them to Huck's, a funky little bar and restaurant just a few blocks away with a deck overlooking a canal off the West Pascagoula.

At Huck's we met Jay Mengle and his wife, Sandy. Mengle, who calls himself Cap'n Jay, is the president of the Pascagoula River Basin Alliance and a river tour guide who takes passengers on sightseeing trips on the Pascagoula. The Mengles live near the river, and Cap'n Jay has developed a love for the Pascagoula and for showing it to others in his 19-foot Caribiana Skiff. He told us they had come to the area three years before, having spent quite some time after his retirement from the military cruising around the eastern half of North America in a motor yacht. Since settling here near the Pascagoula River, Cap'n Jay, an avid birder and naturalist, had probably learned as much about the river and its unique plants and animals as most of the locals. He spent as much time as possible out there cruising the marshes and bayous, and enthusiastically told us about some of his favorite haunts.

Huck's was the perfect ending for a journey down the Pascagoula. We sat out on a deck overlooking a canal off the west river, easing the transition back to civilization with cold beer, burgers, and conversation about our various past adventures. Talking with this group of people who cared about this river so passionately made me realize how lucky we were to be able to experience it in its entirety. I had been a bit skeptical at first when Ernest approached me with the idea for the trip and this book, but sitting there at Huck's after 2 weeks of paddling, I realized how glad I was that I had agreed to make the journey. Mississippi is blessed to have a river system like the Pascagoula. I had just paddled 240 miles of uninterrupted, downstream waterway. Not once did I have to get out of my boat to portage over a dam, go through a lock, or even drag over a fallen tree. For two weeks I had been surrounded by forest, passing only minimal signs of human alteration and civilization in the form of bridges and infrequent dwellings. The Pascagoula River basin is still a place where nature dominates, and the lands surrounding the river and its tributaries are a haven not only for their diverse plant and animal communities, but for all who care to explore them. This river system with all its fine tributaries offers more miles of paddling than most canoers and kayakers would ever want to do. But even those of us like Ernest Herndon, Travis Easley, and me, who have done the system from top to bottom, will likely return for more the next time we get the urge to pick up the paddle and go see a wild river.

AFTERWORD

Amazingly, Scott, Travis and I survived—not only the river but each other. I accepted Scott's inexplicable choice in subpar river vessels, and selflessly did whatever I could to compensate, such as carry his food and boulder-sized sandstone mementoes. I overlooked the fact that Travis did not unfailingly defer to my age, experience, and wisdom. It's even possible, I suppose, that Scott and Travis had to tolerate certain traits of mine, though I can't imagine what. Likewise, we endured the difficulties and downsides of the Pascagoula River system, finding far more to celebrate. I can tick off the irritants: sewer lagoons, litter, cutovers, erosion. More memorable were the beauties: bend after bend of woods, water, and wildlife.

While the Pascagoula is unique from an ecological point of view, as Cathy Shropshire writes in this book's foreword, "the Pascagoula is just one of Mississippi's treasures." I would add that it's just one of the Deep South's treasures. I'm thinking of rivers like Florida's Suwannee and Apalachicola, Alabama's Choctawhatchee and Tombigbee, Mississippi's Pearl and Big Black, Louisiana's Atchafalaya and Ouachita, and east Texas's Neches and Sabine, among others.

The majesties of the Pascagoula open our eyes to the beauties all around us: the wonders of cypress swamps, sandbars, river bluffs, marsh, and woodlands; of wildlife from the lowliest barn swallow to the soaring osprey; of plants from ground-crawling dewberry to sky-scraping loblolly.

Though I sometimes feel world-weary when it comes to canoeing, the Pascagoula inspires me, makes me want to see more, to experience the glories of nature and wildlands that we tend to take for granted. Despite the Pascagoula's ecological distinction, few outside south Mississippi seem to have heard of it. Ideally, it should share in the fame of the Everglades, the Okefenokee, the Atchafalaya, the Big Thicket. But maybe it doesn't need fame. What it needs—and has, to large degree—is protection. Thanks to the farsightedness of some pioneering conservationists/environmentalists, the

Pascagoula is well armored by public lands. Particularly on the main river itself, there are few places where state wildlife management area doesn't border one side or the other, if not both.

More public awareness can help protect the river. The next time some economic developer or government official wants to dam a tributary, dredge a ship channel, or harness natural flooding with levees, people should be able to articulate why it's important for the river to be in its natural state—and why we're proud that it is.

Ernest Herndon

BIBLIOGRAPHY

Brockman, C. Frank. *Trees of North America*. Golden Press, 1986.

Brown, Tom. *Tom Brown's Guide to Edible and Medicinal Plants*. Berkley Books, 1985.

Cain, Cyril Edward. *Four Centuries on the Pascagoula*, vols. 1 and 2. The Reprint Co., 1953, 1962.

Foreman, Dave, and Howie Wolke. *The Big Outside: A Descriptive Inventory of the Big Wilderness Areas of the United States*. Revised edition. Harmony Books, 1992.

Herndon, Ernest. *Canoeing Mississippi*. University Press of Mississippi, 2001.

Jackson, W. Harvell. *By the Rivers of Water: History of George County, Mississippi*, vols. 1 and 2. Lewis Printing Services, Pascagoula, Mississippi, 1978, 1982.

Odenwald, Neil, and James Turner. *Identification, Selection and Use of Southern Plants for Landscape Design*. Claitor's Publishing Division, Baton Rouge, Louisiana, 1996.

Peterson, Lee Allen. *Edible Wild Plants*. The Peterson field guide series, no. 23. Houghton Mifflin Company, 1977.

Romans, Bernard. *A Concise Natural History of East and West Florida*. University of Alabama Press, 1999.

Schueler, Donald G. *Preserving the Pascagoula*. University Press of Mississippi, 1980.

Wells, Mary Ann. *Native Land*. University Press of Mississippi, 1994.

Williams, Scott B. *Exploring Coastal Mississippi*. University Press of Mississippi, 2004.